READ THIS BEFORE IMPLEMENTING

SAP

The No-Nonsense Guide to SAP Success
Without the Expensive Setbacks

SANJJEEV K SINGH

RIVER GROVE
BOOKS

Published by River Grove Books
Austin, TX
www.rivergrovebooks.com

Distributed by River Grove Books

Design and composition by Greenleaf Book Group and Sheila Parr
Cover design by Greenleaf Book Group and Sheila Parr
Cover images used under license from Adobe Stock: 689466591 / Icons-Studio
and Shutterstock: 2488089279 / Yodpol

Publisher's Cataloging-in-Publication data is available.

Paperback ISBN: 979-8-90052-021-6

Hardcover ISBN: 979-8-90052-059-9

eBook ISBN: 979-8-90052-022-3

First Edition

Contents

Introduction

Implementing SAP is unlike any other IT project. It stands as one of the most profound and far-reaching business decisions a company can make—a strategic move that reverberates through every layer of an organization. Its influence extends beyond merely updating systems and streamlining processes; it fundamentally reshapes culture, empowers people, and dictates long-term competitive advantage.

Companies rarely embark on this monumental journey with a casual disposition, acutely aware that the financial investment is staggering, and the operational risks are extraordinarily high. Yet despite this pervasive understanding of the stakes, an unsettling number of SAP projects falter, falling short of their ambitious goals. They frequently balloon over budget, stretch timelines far beyond initial projections, and, most critically, fail to deliver the transformative change they promised.

Why is this a recurring narrative in the corporate world? It's not an indictment of SAP's inherent capabilities; the software itself is a robust, world-class enterprise solution. Nor does the blame lie with a dearth of intelligent individuals or highly capable consultants. The truth, as I have come to understand it over decades, as a former SAP executive and now CEO of ASAR, a trusted SAP partner, is both simpler and profoundly more sobering: A significant portion of SAP projects are destined for disappointment long before the first line of code is configured or the initial system setting is tweaked. A company's fate is often sealed in the formative stages—in the nascent vision, in the preparatory groundwork, and, crucially, in the critical conversations that executives either embrace or deliberately sidestep, months or even years before the actual implementation phase begins.

It is precisely this unsettling reality that compelled me to write this book. My professional life has been dedicated to the intricate world of digital transformation and, more specifically, SAP implementations. Over the past several decades, I have been a privileged witness to both spectacular successes and agonizing failures. I have observed companies execute radical shifts in their business models, elevate customer experiences to unprecedented heights, and achieve operational efficiencies that once seemed unattainable.

Conversely, I have also seen organizations become paralyzed by indecision and overwhelmed by complexity or left deeply disillusioned by prohibitively expensive projects that never materialized into their envisioned state.

Through the crucible of these diverse experiences—the triumphs and the setbacks—one irrefutable truth has emerged: The ultimate success of an SAP implementation has strikingly little to do with the underlying technology itself. Instead, it is inextricably linked to the caliber of leadership, the unwavering clarity of purpose, and the disciplined execution brought to bear on the project.

To truly grasp the strategic significance of SAP today, it's helpful to understand the journey of enterprise resource planning (ERP) itself. Born from the need to integrate disparate business functions, early ERP systems aimed to centralize data and streamline operations. Initially focused on manufacturing resource planning (MRP), ERP evolved to encompass finance, human resources, and supply chain.

SAP, founded in 1972, has been a central figure in this evolution. From its mainframe roots with R/2 to the client-server architecture of R/3, and now to the in-memory computing power of S/4HANA, SAP has continually adapted its technology to meet the changing demands of global business.[1] This long history and constant innovation have solidified SAP's position as a market leader—not merely as a software vendor but as a standard setter for how complex enterprises can manage their operations.

The move to S/4HANA in particular isn't just an upgrade; it's a fundamental shift enabled by new database technology (SAP HANA). This allows for real-time analytics, simplified data models, and the potential for intelligent automation (AI/ML). This evolution reinforces the idea that SAP is not a static tool but a dynamic platform for continuous business innovation. Understanding this trajectory helps leaders appreciate that an SAP implementation isn't just about "getting current" but about positioning the organization for future capabilities and competitive advantage.

This book is specifically crafted for the leaders who stand at the precipice of this transformative journey. Whether you hold the mantle of CEO, CFO, CIO, or any senior business executive sponsoring an SAP project, my singular objective is to equip you with the insights necessary to circumvent the most pervasive pitfalls, to prepare you for the multifaceted challenges and opportunities that genuinely lie ahead, and to empower you to approach implementation with a profound, unblinkered understanding of its implications.

You will not find granular technical configuration instructions within these pages, nor will you encounter screenshots detailing system settings. Instead, this is a repository of strategic guidance, hard-won practical lessons,

---

1 "The History of SAP," SAP, accessed October 14, 2025, https://www.sap.com/about/company/history.html.

and candid stories from the trenches—all designed to help you adeptly navigate one of the most critical business transformations your organization will ever embark upon.

The pervasive misconception that undermines countless SAP initiatives is the tendency to approach them as a mere technical upgrade. This narrow perspective views SAP as simply a replacement for outdated software—a newer, shinier version of an existing system. The prevailing logic often articulates something akin to: "Our legacy ERP system is obsolete; migrating to SAP S/4HANA will grant us enhanced speed, contemporary features, and a modernized technological backbone. Once we successfully implement it, our current operational dilemmas will, by extension, be resolved."

This fundamental misunderstanding is precisely why this book is indispensable. Because before you even contemplate the intricate details of implementing SAP, you must first cultivate a profound understanding of the monumental undertaking you are genuinely committing to. It's about grasping the difference between installing new software and redesigning your entire organizational circulatory system.

This seemingly logical mindset is, in fact, a deeply insidious trap. SAP is emphatically not just a software package; it is, at its core, a comprehensive operating model for your entire business. To truly leverage its power, an implementation necessitates a fundamental reimagining of how your organization operates at every level. This includes, but is not limited to, how you approach sales, manage manufacturing processes, deliver customer service, oversee financial operations, and, critically, how strategic decisions are formulated and executed. SAP permeates every functional domain, every department, and every corner of your enterprise. If you persist in treating this profound overhaul as a mere IT project, you will inevitably squander the immense transformation opportunity it presents, and, ironically, likely create a new array of complex problems that far outweigh the ones you sought to solve.

An SAP project is undeniably a substantial financial investment—a fact that cannot be understated. The direct costs alone—encompassing software licenses, external implementation services, and the allocation of internal human resources—can effortlessly escalate into the millions, frequently climbing into the tens of millions, of dollars. For multinational corporations, these figures can easily reach hundreds of millions. However, to focus solely on these monetary outlays is to miss the far greater picture; the true stakes involved are considerably higher and far more existential.

Every SAP project, by its very nature, places your business continuity, your meticulously cultivated customer relationships, and the daily productivity of your entire workforce at profound risk. A poorly conceived or mismanaged implementation can trigger a cascading series of detrimental effects: Your supply chain might grind to a halt, customer orders could experience unacceptable delays, and your employees could be left bewildered and struggling to perform even the most rudimentary tasks within unfamiliar, unintuitive, or broken systems. The ripple effects of such disruptions can undermine market confidence, damage brand reputation, and lead to significant talent drain.

Conversely, a meticulously planned and successfully executed SAP project can transcend its operational role to become a powerful engine for sustained organizational growth and innovation. It possesses the unique capacity to unify disparate and fragmented operations under a single, cohesive framework. It can furnish leaders with invaluable, real-time insights, enabling agile decision-making in increasingly volatile and fast-changing global markets. More fundamentally, it can empower your teams with processes that are inherently simpler, dramatically faster, and significantly more reliable, thereby fostering a culture of efficiency and precision. Such a transformation can unlock your organization's capacity to scale rapidly, expand seamlessly into new global territories, and innovate in ways that your archaic legacy systems would have perpetually stifled.

In essence: The consequences of failure can set your organization back years, potentially crippling its competitive posture for a generation. Conversely, the dividends of success can propel your enterprise forward, establishing a formidable foundation for growth and innovation that endures for decades.

It is for these reasons that the decisions you make before the project officially commences—the meticulousness of your preparation, the discernment with which you select your partners, and the proactive mindset you adopt—are, unequivocally, the most pivotal decisions you will make throughout the entire journey.

Throughout this book, we will systematically examine each of these pervasive myths, providing you with practical frameworks and actionable strategies to replace them with a mindset that is demonstrably geared toward success.

The pages that follow are not theoretical treatises; they are infused with the raw realities of real-world SAP projects. You will encounter compelling stories from companies—some that navigated the complexities to achieve brilliant success, others that nearly crumbled under the sheer weight of their own missteps. These narratives are not abstract case studies; they are drawn directly from decades of my personal experience collaborating with organizations of every conceivable size and across a diverse spectrum of industries, from small regional manufacturers to global conglomerates.

While this book is thoughtfully structured into distinct chapters for ease of comprehension, it is intentionally not divided into separate "parts" catering to different audiences. Instead, it is conceived as a continuous narrative, mirroring the natural, evolving flow of your SAP journey as a leader. Think of each chapter as one of the following stepping stones on a singular, unified path:

- **Laying the foundation:** This initial phase delves into the critical groundwork—defining a compelling vision, crafting a robust strategy, ensuring meticulous organizational alignment, and establishing the imperative for impeccable data quality and governance. It's about setting the stage for success.

- **Making critical choices:** Here, the focus shifts to the pivotal decisions that will profoundly shape your project's trajectory—selecting the optimal implementation partners, meticulously defining the scope, discerning when to customize and when to standardize, and establishing clear, actionable priorities for phased deployment.

- **Leading transformation:** This section addresses the active, ongoing leadership required to navigate the complexities of change—guiding your people through significant shifts, steadfastly avoiding counterproductive shortcuts, and meticulously building a resilient and sustainable platform for tomorrow's business needs.

- **Sustaining success:** The final phase extends beyond the go-live, focusing on embedding the hard-won lessons learned, continually strengthening organizational capabilities, and cultivating a culture of iterative improvement, ensuring your SAP system and the processes it enables remain optimized and continue to deliver value long after initial deployment.

I did not undertake the arduous task of writing this book with any intention of merely showcasing technical knowledge or demonstrating a mastery of SAP's intricate functionalities. My motivation is disarmingly simple yet powerfully deep: I want more companies to experience that profound kind of success and, conversely, far fewer to endure the agonizing pain, financial drain, and demoralization that accompanies failure. This stems from a deeply personal imperative: I have witnessed, with disheartening frequency, far too many organizations stumble, falter, and even catastrophically fail on this immensely challenging yet profoundly rewarding journey.

I have spent countless hours in executive boardrooms observing heated arguments over project scope while the foundational vision remained dangerously undefined and ignored. I have watched, with a sense of growing despair, highly capable teams burn themselves out under the relentless pressure of endless, often valueless, change requests that bloated project timelines and budgets. I have seen esteemed consultants walk away from engagements, leaving their clients saddled with newly implemented SAP systems so profoundly over-customized that they became virtually impossible to support, maintain, or upgrade, turning a supposed asset into a crippling liability.

Yet, amid these disheartening observations, I have also been privileged to witness the antithesis: organizations that approached SAP not as a burden but as a catalytic opportunity. These were companies that viewed the implementation as a chance to fundamentally rethink their future, to strategically align their diverse teams, and to ruthlessly simplify their core business processes. The projects undertaken by these enlightened organizations were not merely successful in a technical sense—they were genuinely transformative, unlocking unprecedented levels of efficiency, agility, and competitive advantage.

As you embark on reading this book, I earnestly encourage you to approach its contents with a spirit of humility and an insatiable curiosity. Resist the comfortable, yet dangerous, assumption that "this won't happen to us." Recognize, with clear-eyed pragmatism, that the inherent risks associated with an SAP implementation are unequivocally real and substantial. But simultaneously embrace the understanding that the opportunities for profound, lasting business transformation are equally real and exponentially rewarding.

Think of this not as a dry, technical manual, but rather as a living, breathing leadership guide. Because at the very core of it, SAP is not implemented by lines of code, complex algorithms, or sophisticated hardware; it is ultimately implemented by people—your people, your partners, and your customers. And among these, the people who matter most, whose influence will be most determinative, are you, the leaders. You are the ones who must courageously chart the course; make the tough, often unpopular, calls; and, most critically, inspire unwavering confidence and commitment within your teams as they navigate uncharted territory.

If you distill nothing else from this comprehensive introduction, let this singular truth resonate deeply within you: Your SAP project will ultimately succeed or fail not because of the software itself but because of the quality, foresight, and unwavering dedication of the leadership you bring to it.

This book is here, unequivocally, to help you bring that essential leadership to vivid, transformative life.

The Ideal SAP Implementation Process Overview

Implementing SAP is frequently, and aptly, described as embarking on a "journey." However, without a meticulously crafted map, a clear destination, and a reliable compass, even the most promising journeys can quickly devolve into costly and frustrating detours. All too often, companies, driven by an understandable eagerness to move swiftly, plunge into these projects without fully grasping the intricate, multifaceted stages of transformation that lie ahead. The outcome is regrettably predictable: significant delays, escalating frustration, budget overruns, and a painful failure to meet the lofty expectations initially set.

It is precisely to counteract this common pitfall that it becomes paramount to begin by developing a profound understanding of what the ideal SAP implementation process truly entails. Taking the time to gain a high-level, comprehensive bird's-eye view of the entire life cycle will provide invaluable foresight in maintaining momentum and making informed choices. By tracing the journey from the pivotal moment your leadership team makes the strategic decision to move forward with SAP—through the complex stages of design and build to the critical day your new system goes live, and well beyond, into the crucial period of stabilization and continuous improvement—you begin to see that true transformation is not a single event but an ongoing evolution of people, processes, and technology working in harmony.

While the landscape of SAP implementation methodologies offers various approaches—from SAP Activate, with its structured road maps and accelerators, to highly agile, iterative frameworks, and even hybrid models blending both—the underlying, immutable logic of a truly successful SAP project consistently adheres to six fundamental, sequential, and deeply interconnected phases. These phases represent a logical progression of activities, each building upon the successful completion of the previous one, ensuring a cohesive and well-managed transformation. Let's look at the six phases that make an ideal SAP project implementation. Each phase is briefly introduced here, and the following chapters will explore them in greater depth, defining your road map to a successful SAP project.

PHASE 1: VISION AND STRATEGY ALIGNMENT— LAYING THE STRATEGIC CORNERSTONE

Every genuinely successful SAP project begins not with the technical specifications of software, nor with complex system architecture diagrams, but rather with a profound and crystal clear articulation of business strategy. This is the phase where the executive leadership team establishes the fundamental tone, direction, and overarching purpose for the entire transformation. Without this strategic bedrock, even the most technically brilliant implementation will lack direction and struggle to deliver meaningful business value.

Before any discussion of system capabilities or project timelines, the leadership team must unequivocally define the fundamental "why." Why are you embarking on this colossal SAP implementation now? What are the profound business challenges or compelling opportunities that necessitate such a significant investment and transformation?

The precise answer to this "why" is not a mere rhetorical exercise; it will profoundly shape and inform every single decision that follows, from scope definition and budget allocation to partner selection and change management strategies. A project driven solely by the desire for "new technology" or "catching up" often lacks the strategic depth to overcome inevitable hurdles.

Next is establishing a North Star, which brings unity to the project's vision. Projects, especially those as large and complex as an SAP implementation, are highly susceptible to fragmentation and scope creep when there is no singular, unifying vision—a North Star—to guide the myriad of decisions that arise. A clear, concisely articulated strategic objective acts as an indispensable guiding principle, particularly when difficult trade-offs must be made between competing business priorities, departmental demands, or technical feasibility.

This North Star provides a nonnegotiable anchor, ensuring that individual departmental needs or technical preferences do not inadvertently derail the overarching strategic intent. It empowers project teams to make aligned decisions and gives the steering committee a clear criterion for resolving conflicts.

The vision for an SAP transformation cannot, under any circumstances, be outsourced. While external consultants bring invaluable expertise in refining, articulating, and translating that vision into a feasible plan, the ultimate ownership and stewardship must unequivocally reside with the executive leadership team. Without genuine, sustained, and visible leadership alignment and sponsorship at this foundational stage, the project is inherently compromised, essentially beginning off course before the first step is even taken.

Companies that mistakenly bypass or superficially address this foundational phase often confuse intense activity with genuine progress. They may rush directly into software selection or design workshops, only to find themselves perpetually debating fundamental objectives, struggling with stakeholder buy-in, and experiencing constant scope churn. The result is typically a project that lacks purpose, suffers from internal resistance, and ultimately fails to deliver the promised strategic value, regardless of technical execution. It becomes a technology project looking for a business problem to solve rather than a strategic transformation leveraging technology.

PHASE 2: PREPARATION AND DATA READINESS—FORTIFYING YOUR FOUNDATION

If strategy and vision represent the unshakable foundation of your SAP transformation, then data is unquestionably the primary building material. And much like constructing a physical edifice, using shoddy or unreliable materials will inevitably result in a weak, unstable, and ultimately failing structure. This phase is arguably the most underestimated and often neglected, yet its successful execution is absolutely critical.

The vast majority of companies harbor a significant, and often shocking, underestimation of the true state and quality of their existing data. Decades of disparate systems, manual processes, and inconsistent data entry practices typically leave organizations with a complex "data mess." This often includes

- Inconsistent customer names (e.g., "IBM Corp," "International Business Machines," "IBM") making unified customer views impossible

- Duplicated product codes or material masters, leading to inventory discrepancies and purchasing errors

- Unreliable inventory counts or bill of materials, impacting production and customer fulfillment

Those are just a few of the "data messes" that can be spotted. Attempting to simply migrate this pervasive "digital chaos" into a sophisticated, integrated SAP system will not miraculously resolve these issues; it will, in fact, merely digitize and amplify the chaos, leading to immediate operational problems, inaccurate reporting, and a profound erosion of user trust in the new system. The mantra here must be "garbage in, garbage out." A rigorous, detailed master data assessment is essential to understand the current state, identify critical data quality gaps, and quantify the effort required for remediation.

Given the critical role of data, establishing a robust Data Governance Framework is nonnegotiable. This framework defines the rules, roles, and responsibilities for managing data throughout its life cycle.

Without clearly defined accountability and established processes for data creation, maintenance, and quality checks, data will inevitably deteriorate post-go-live, undermining the very foundation of your new SAP system.

Preparation extends far beyond technology and data; it crucially involves preparing your people. An SAP implementation is, at its heart, an organizational transformation, and neglecting the human element is a common cause of failure.

This phase determines whether your SAP project will stand tall, supported by an organization ready to embrace new ways of working, or whether it will collapse under the weight of poor data and human resistance. It's about building confidence and buy-in before the changes are upon them.

PHASE 3: DESIGN AND BLUEPRINTING— TRANSLATING VISION INTO ARCHITECTURE

This is the pivotal phase where your high-level strategic vision begins its intricate translation into a tangible, detailed future-state business model, solution design, and technical architecture. The design phase is not about producing reams of impenetrable documentation; rather, it is about engaging in a disciplined, iterative process of making hard, informed choices that align with your North Star and prepare for sustainable operations.

A critical paradigm shift is required in this phase. Instead of beginning with the question "How can SAP be configured to precisely replicate our existing, often inefficient, legacy process?" the more strategic and beneficial question is "How can our current business process be optimized and aligned with SAP's industry best practices and standard functionalities?"

These Fit-to-Standard workshops are collaborative sessions involving business process owners, key users, and SAP functional consultants. This is perhaps the most significant battleground of the design phase. Every single deviation from SAP's standard functionality—whether it's a minor configuration change or a major custom development (often referred to as RICEFW: reports, interfaces, conversions, enhancements, forms/workflows)—introduces additional cost, elevates risk, and increases the long-term total cost of ownership (TCO).

A disciplined approach demands that any proposed customization must be rigorously justified based on a clear business imperative that cannot be met by standard functionality or process reengineering. The default answer should always be no to customization, unless an overwhelming strategic or competitive advantage can be demonstrated. Simplicity scales; complexity kills.

The design process must extend beyond merely solving today's immediate pain points. A truly ideal SAP implementation considers the trajectory of the business, actively future-proofing the design. This involves asking critical questions that we will explore deeper in Chapter 8.

Good design is characterized by clarity, robustness, and adaptability. Poor design, by contrast, is often the result of endless compromises, indecision, or a myopic focus on current challenges, leading to a brittle system that quickly becomes obsolete.

PHASE 4: BUILD AND TEST—
BRINGING THE SYSTEM TO LIFE

This is the phase where technology finally moves into the primary spotlight. The designs conceptualized in the previous phase are now meticulously translated into a tangible, configured, and often custom-developed SAP system. However, even here, success is ultimately more about disciplined execution and rigorous validation than it is about simply writing code.

The era of monolithic "big bang" builds stretching for years is largely a relic of the past. Modern, ideal SAP implementations increasingly adopt agile principles, even within structured methodologies like SAP Activate. This means building in short, focused cycles, with two-to-four-week sprints with continuous feedback and iterative refinement.

This iterative approach drastically reduces the risk of building the wrong solution, fosters stronger collaboration between technical and business teams, and significantly improves user acceptance by involving them throughout the development process. Long, opaque "big bang" builds almost invariably disappoint because they delay user engagement and feedback until it's too late for cost-effective changes.

The actual migration of data from legacy systems into SAP is a complex, high-stakes activity that often overlaps with the build phase. It requires mapping legacy data fields to new SAP fields and actively cleaning and enriching data before loading (as identified in Phase 2). Often, multiple "dry runs" or "mock migrations" are performed to identify issues and refine the process before the final go-live cutover.

That is why various types of testing are crucial for validating processes, not just functions. Testing is not a one-time event; it's a continuous, multi-layered activity throughout the build phase. And critically, it's not just about whether the system works technically; it's about whether the process makes sense for real users and whether the system supports the end-to-end business flow. This involves unit, integration, user acceptance, performance, security, and regression testing.

Rigorous testing requires realistic test data, well-defined test cases, and a dedicated team of business users and testers. Neglecting any of these testing stages is a common path to post-go-live chaos.

As technology continues to move into the spotlight during this phase, it's equally important to remember that it's a balance of speed and direction. In other words, while agility emphasizes speed, it's crucial to balance rapid execution with unwavering adherence to the strategic direction. Moving fast is meaningless—and potentially dangerous—if you are configuring or developing the wrong thing. This balance is maintained through regular sprint reviews and demos to ensure alignment with business expectations; empowered product owners to make quick, informed decisions on scope and priorities; and a clear definition of "done," so all team members understand what constitutes a completed and tested piece of work.

A small correction or pivot in this phase, informed by early testing and feedback, prevents exponentially larger rework and cost overruns later in the project. The best implementations are not necessarily the fastest builds but the most disciplined and iteratively aligned ones.

PHASE 5: DEPLOY AND GO-LIVE—
THE MOMENT OF TRUTH

Go-live is, without a doubt, the most visible, high-pressure, and often most misunderstood moment of the entire SAP project. It's the point where months or years of planning, preparation, design, and build culminate in the switch from old systems to the new SAP environment. Crucially, go-live is not the finish line; it is the starting line for your new business reality.

The transition from legacy systems to SAP is inherently a high-wire act, requiring meticulous planning and flawless execution. Every single open transaction, every financial balance, every customer order, every shipment currently in transit, and every piece of inventory must be meticulously accounted for and correctly transitioned into the new system. There are, realistically, no second chances to get this right without significant business disruption.

No matter how thoroughly you prepare, the first day, week, or even month operating in the new SAP system will feel foreign, perhaps even disorienting, to most employees. Muscle memory built over years with legacy systems doesn't disappear overnight. For users to effectively utilize and ultimately trust the new system, it requires a foundation of confidence built over months of communication, training, and early engagement.

Successful change management at go-live is about anticipating user frustrations and providing immediate, empathetic, and effective solutions.

Success at go-live is not solely measured in the number of transactions posted correctly; it is equally measured in the confidence and affirmation expressed by your people: "Yes, I can confidently do my job in this new system." Effective communication during this critical stage involves both internal and external communication while celebrating milestones along the way.

Go-live signifies the birth of your new operating model, not the cessation of effort.

PHASE 6: STABILIZATION AND CONTINUOUS IMPROVEMENT—REALIZING LASTING VALUE

The true test of your SAP project's success, and indeed its long-term return on investment, materializes not in the euphoria of go-live but in the sustained period that follows—when the initial excitement inevitably fades and the day-to-day realities of operating a new system set in. This phase is crucial for realizing the promised benefits and ensuring the SAP system becomes an enduring asset, not a static cost center.

The first thirty, sixty, or even ninety days after go-live, commonly referred to as *hypercare*, are arguably the most intense period for end users and the support team. It is a highly dynamic phase where issues will undoubtedly arise. These can range from minor user errors to unexpected system performance issues, data discrepancies, or integration glitches. This is not a sign of failure; it is a normal and anticipated part of any large system deployment.

During this phase, productivity may temporarily dip as users adapt to new processes and interfaces; their efficiency may temporarily decrease.

The critical factor is not whether issues arise but whether you are prepared to identify, prioritize, and resolve them swiftly and effectively. A well-structured hypercare phase, with clear escalation paths and robust communication, transforms initial challenges into learning opportunities and builds user confidence.

To truly embed the new ways of working, change needs to stick. Old habits are notoriously difficult to break. Without active reinforcement and consistent leadership, employees may inadvertently, or even deliberately, attempt to "work around" the new SAP system, reverting to inefficient legacy processes or creating manual work-arounds. Organizations must consider ongoing training and refreshers and performance monitoring with key performance indicators (KPIs), and leaders must consistently model commitment to the new processes.

This sustained focus ensures that the new system truly becomes the ingrained standard for daily operations, maximizing its value.

The ideal implementation process does not merely aim for a one-time success at go-live; it actively builds the organizational capability for ongoing evolution and continuous improvement. An SAP system, in an ideal scenario, is never truly "finished." The business environment is dynamic, and technology continues to evolve.

This phase is often ignored in the post-go-live exhaustion, but it is precisely where the long-term return on investment (ROI) of your massive SAP investment is either fully realized or, sadly, squandered. It's about cultivating a culture of perpetual optimization.

YOUR STRATEGIC COMPASS

While the six phases provide a structural road map, the overarching role of leadership is not confined to any single phase; it is a continuous, pervasive force that runs across and binds all of them. It is easy to dissect the SAP implementation process in abstract, theoretical terms. But how does an ideal project actually feel from the perspective of the people immersed in it—the core project team, the business users, and the executive sponsors? The qualitative experience of the project team often serves as a powerful leading indicator of ultimate success.

An ideal SAP project meets the following criteria:

- Meetings are purposeful, crisp, and conclusive, not chaotic or endless. They have clear agendas, empowered decision-makers, and actionable outcomes.

- Decisions are made with consistent reference to the defined strategic vision and the North Star, not swayed by internal politics, historical biases, or individual preferences.

- External consultants serve as expert guides, coaches, and implementers, but the executive team and internal business leaders remain unequivocally in the driver's seat. Ownership is clearly internal.

- Employees across all levels feel genuinely supported, informed, and empowered to adapt to the changes rather than feeling blindsided, overwhelmed, or resentful. There is a sense of shared journey.

- By the time go-live approaches, there is a palpable sense of confidence, readiness, and collective anticipation rather than widespread panic, last-minute scrambling, or anxious uncertainty.

- Communication is open, honest, and frequent, acknowledging challenges while celebrating progress.

- Mistakes are viewed as learning opportunities, not as reasons for blame.

- Cross-functional collaboration is the norm, not the exception.

- There is a visible enthusiasm for the future-state capabilities and the benefits the new system will bring.

Of course, not every project follows the ideal, disciplined path outlined earlier. The tempting allure of cutting corners, skipping phases, or underinvesting in critical areas—often driven by perceived time or cost pressures—almost invariably leads to a cascade of negative consequences. The cost of getting an SAP implementation wrong extends far beyond the initial financial investment; it is exponential and often inflicts lasting damage across the organization.

The journey to an ideal SAP implementation may be arduous, demanding significant effort, investment, and resilience. However, the lasting rewards, in terms of operational excellence, strategic agility, and competitive advantage, are profound and sustainable for decades to come. In essence, without rigorous discipline, strategic foresight, and unwavering leadership, the cost of failure is not merely financial—it is deeply cultural, operational, and ultimately, a significant impediment to future growth and innovation.

The concept of an "ideal" SAP implementation process is not a utopian fantasy or an unattainable theoretical construct. On the contrary, it represents a tested, repeatable, and fundamentally sound journey that thousands of successful companies across the globe have meticulously navigated. It is about deliberately setting expectations: An SAP project is fundamentally not a straight, easy road; it is, more accurately, a demanding, often challenging but ultimately rewarding climb. But just as any formidable mountain climb is conquered step-by-painstaking-step, with a clear map and a reliable compass, you can navigate this ascent without wasting precious energy on unnecessary detours or losing your way.

The paramount question, therefore, is not whether such a path exists but whether you, as a leader, will approach this monumental endeavor with the requisite vision, the meticulous preparation, the unwavering discipline, and the consistent leadership required to successfully follow that proven path.

In the chapters that follow, we will dive significantly deeper into each element touched upon here. But before you delve into those granular details, I urge you to hold onto this crucial big picture. Because when the project inevitably feels overwhelming, complex, or fraught with challenges—and it will—this comprehensive overview will serve as your steadfast anchor, reminding you precisely where you are in the journey, what the next logical steps entail, and, most importantly, why this transformative climb is unequivocally worth every effort.

Don't Outsource Your Vision

When organizations embark on an SAP implementation project, the temptation to lean heavily on external consultants is incredibly strong. It's an understandable impulse: After all, these consulting firms and their experienced professionals have navigated dozens, perhaps hundreds, of similar projects before. They possess deep expertise in the methodologies, are intimately familiar with the intricate milestones, and wield a mastery of the tools required for such a complex undertaking. Consequently, senior leaders often fall into the trap of thinking, *They've seen this movie before—they know the script and the ending. Let's simply let them drive the entire initiative.*

However, this seemingly logical delegation is, in my experience, the single greatest and most profound mistake executives make in the realm of digital transformation, particularly with SAP. You can, and indeed should, outsource specialized technical expertise. You can delegate the intricate configuration of the software. You can even contract out the custom development work required to meet unique business needs. But there is one fundamental, indispensable element that you cannot, under any circumstances, outsource: your vision.

Because when you relinquish your vision to external parties, you are not merely delegating work; you are, far more dangerously, abdicating core leadership responsibility. An SAP project, stripped of this vital internal leadership and a clear, owned vision, becomes akin to a majestic ship embarking on a perilous journey without a captain at its helm. It may possess powerful engines and a skilled crew, but without direction, it will inevitably drift aimlessly, buffeted by competing currents, until it ultimately founders upon the rocks of missed objectives, inflated costs, and internal disillusionment.

Too often, companies mistakenly conflate "vision" in the context of an SAP implementation with a generic business case document or a lengthy, detailed list of functional and technical requirements. While these documents are certainly necessary components of a well-run project, they are not the vision itself. Vision, in the context of an SAP transformation, is far more profound, strategic, and foundational. It is the unwavering North Star that provides the ultimate clarity and direction for every single decision, every trade-off, and every resource allocation throughout the multiyear journey.

True vision unequivocally defines several key elements of an SAP implementation:

- The existential "why": Beyond just replacing old software, why does this monumental project genuinely exist for your business? Is it a strategic imperative to unify fragmented systems across global operations? Is it to enable rapid scaling into dynamic new markets or support an aggressive mergers and acquisitions strategy? Is the primary driver to radically improve customer experience, perhaps by streamlining order-to-cash processes or enhancing real-time service capabilities? Or is it about fundamentally shifting to a digital-first business model, requiring agile, data-driven insights at every turn? The answer to this deep "why" shapes every subsequent design choice and prioritization.

- The future-state landscape: What will your business look and feel like after the transformation? This isn't about specific screen layouts; it's about strategic outcomes. Will you achieve a single, undisputed version of truth in finance that empowers real-time analytics for executive decision-making? Will you establish a streamlined, touch-less quote-to-cash cycle that delights customers and accelerates revenue? Will you implement a global standard template across all entities, ensuring consistency, efficiency, and scalability from the factory floor to the boardroom? This paints a vivid picture of the desired future operational reality.

- The transformed operating model: How will the business fundamentally operate differently—not just faster, but in entirely new, more effective ways? This transcends mere process optimization. It's about creating an integrated enterprise where people, data, and decisions are seamlessly aligned. It might mean shifting from reactive problem-solving to proactive forecasting, enabling self-service for customers, or empowering employees with integrated data previously siloed in disparate systems. It's about the cultural and operational shifts that unlock new levels of performance.

Crucially, vision is not technical. It is not concerned with the granular detail of whether a specific data field should be twenty or forty characters long or which particular SAP module will handle a specific transaction. Vision is about business transformation at the highest, most strategic level. It addresses the fundamental aspirations and strategic imperatives of the enterprise.

If this profound vision is not unequivocally defined—and, more importantly, owned and championed—by the executive leadership team, then regardless of the unparalleled talent, experience, or methodology brought by external consultants, the SAP project will inevitably devolve into a chaotic collection of disconnected activities, lacking any unifying strategic thread. It becomes a series of tactical exercises rather than a cohesive, purposeful journey.

WHY COMPANIES TRY TO OUTSOURCE IT: THE ALLURE OF DELEGATION

Given the profound importance of vision, why do so many intelligent and capable leaders succumb to the temptation of outsourcing it? The reasons, while ultimately misguided, are often understandable, rooted in common organizational dynamics and pressures.

One factor is what might be called the "consultants speak the language" fallacy. Many internal leaders, despite their deep business acumen, feel a perceived lack of specific SAP knowledge or prior implementation experience. They believe they don't possess the technical lexicon or architectural understanding to confidently define the vision themselves. Consequently, they assume that consultants, with their specialized jargon and project histories, inherently "know better" and can articulate the necessary vision more effectively. This leads to a passive acceptance of consultant-driven ideas rather than a critical assessment against internal strategic goals.

Another factor is the pressure for speed over clarity. In today's fast-paced business environment, there's immense pressure to demonstrate progress quickly and deliver tangible results. This urgency often leads executives to bypass the arduous, time-consuming, and often contentious work of internal alignment and vision definition. They may jump straight into detailed design workshops or technical discussions, hoping to accelerate the project, only to find themselves bogged down later in endless debates because the fundamental "why" was never truly solidified. Rushing the vision phase invariably costs more time and money in the long run.

There is also a comfort in delegation, the mindset of "let the experts handle it." Leaders are accustomed to delegating tasks to subject-matter experts. They may rationalize that "SAP implementation isn't our core business; our expertise is in manufacturing, or sales, or finance. Let's leave this complex technology initiative to the experts who do this for a living." While delegating execution is wise, delegating ownership of the strategic intent is akin to hiring an architect and then letting them decide what kind of building you want to live in.

Finally, many leaders avoid internal conflict and tough decisions. Defining a clear, transformative vision for SAP often requires making difficult choices. It might mean standardizing processes that some departments prefer to keep unique or prioritizing one strategic objective over another. These are inherently political and emotionally charged discussions. Delegating the vision to consultants can feel like a convenient way to defer or avoid these challenging internal confrontations, allowing external parties to bear the brunt of unpopular decisions. However, this only postpones the conflict, often exacerbating it later.

While the desire to leverage external expertise is healthy, it is critical to recognize that you can outsource technical execution, but you absolutely cannot outsource ownership of your company's strategic future. Only the internal leadership team possesses the intimate understanding of the business's unique competitive landscape, its core values, its long-term aspirations, and its organizational culture. This invaluable internal knowledge is the bedrock upon which a truly transformative SAP vision must be built.

When the executive leadership abdicates its responsibility for defining and owning the SAP vision, the project inevitably collapses into a painful cycle of tactical firefighting, reactive decision-making, and a chronic lack of direction. The symptoms of an outsourced vision are unmistakable and profoundly damaging. Without a clear, universally understood North Star, every individual department, business unit, or functional leader will naturally default to advocating for their own parochial version of success. Finance will demand rigid controls, Sales will push for maximum flexibility and customization, Operations will prioritize speed and unique workflow adherence, and IT will focus on technical elegance. With no overarching vision to align these competing demands, the SAP project morphs into an exhausting, internal tug-of-war. Decisions become compromises, scope grows uncontrolled, and the project team is paralyzed by conflicting demands, leading to delays and resentment.

In the absence of a strong, guiding strategic direction, external consultants often feel compelled to satisfy every single request from every department, irrespective of its strategic value or alignment with SAP best practices. The default answer becomes "yes" to customization rather than a disciplined "why" or "how can we fit to standard?" This leads to a monstrous "Frankenstein system"—a patchwork of bespoke code, complex interfaces, and nonstandard processes. Such a system is not only astronomically expensive to build but also becomes a perpetual drain of resources for maintenance, support, and future upgrades. It strips away the very benefits of integration and standardization that SAP is designed to provide.

When employees perceive the SAP initiative as a "consultant-driven project" rather than a genuine company transformation, a pervasive sense of detachment and cynicism sets in. They feel the change is being "done to them," not "done with them" or "done for the business." This breeds resistance, hinders user adoption, and results in the new system being underutilized or even actively circumvented. The significant investment in the software yields minimal operational impact if the people who are meant to use it don't believe in its purpose or value.

A project where vision is outsourced may, technically, "go live." The system may be installed and data migrated. However, without a clear strategic purpose guiding its design and implementation, SAP becomes merely a technical replacement for an old system, not a catalyst for genuine business transformation. The organization may have spent millions, but the promised improvements in efficiency, agility, or competitive advantage fail to materialize. The project might be considered "successful" by IT, but it delivers little or no lasting business value.

The indecision, conflicting priorities, and rampant customization born from an outsourced vision directly translate into significant financial and schedule impacts. Endless design cycles, rewrites of requirements, and custom development efforts balloon the budget. Project timelines stretch indefinitely as teams grapple with a lack of clear direction, leading to frustrated stakeholders and a diminishing return on investment. Rework becomes the norm, consuming resources that should be focused on forward progress.

A system implemented without a cohesive vision can introduce new operational risks. If processes are not truly optimized or if data integrity is compromised due to conflicting requirements, the new SAP system can actually hinder daily operations, lead to financial errors, or even impact compliance, creating more problems than it solves.

Outsourcing your SAP vision is the fastest, most expensive, and most damaging way to spend millions on a sophisticated system that, ultimately, fails to deliver genuine business transformation and can even leave the organization in a worse state than before.

WHAT LEADERSHIP OWNERSHIP LOOKS LIKE: THE CAPTAIN AT THE HELM

Owning the SAP vision does not imply that executive leaders must become SAP technical experts or immerse themselves in every granular configuration detail—far from it. Instead, it means they must set the strategic boundaries, define the ultimate destination, and provide the unwavering guidance within which consultants, project teams, and business users can operate effectively. It's about being the captain at the helm, not necessarily the engineer in the engine room.

Strong leadership ownership of the SAP vision manifests through tangible actions and consistent behaviors like the following:

- A crystal clear statement of purpose: The CEO, or the executive sponsor, should be able to articulate, simply and concisely, in a single sentence or a brief paragraph, why the company is investing in SAP.

This statement must be easily understood by every employee, from the executive suite to the front lines. It should be communicated relentlessly—in town halls, internal newsletters, project kickoffs, and informal discussions—serving as the project's mantra.

- Nonnegotiable strategic priorities and design principles: Leaders define what truly matters most for the business, setting clear guardrails for design decisions. These are not merely suggestions but firm principles that guide every choice. Examples include:

  - "Global consistency over local customization unless there's a legal or significant competitive differentiator."

  - "Simplicity and user experience over complexity or replicating legacy inefficiencies."

  - "Data integrity and real-time visibility are paramount for all financial and operational reporting."

  - "Customer experience will always take precedence over internal convenience." These principles act as powerful filters, guiding the project team when conflicts arise and preventing scope creep.

- Visible and consistent engagement: Executives don't just appear for the project kickoff and then disappear until go-live. They demonstrate their commitment through active, visible engagement. This means regularly attending key steering committee meetings, actively participating in critical design review workshops, making appearances at town halls, communicating progress to the broader organization, and modeling the desired behaviors (e.g., actively using early prototypes, asking questions about new processes). Their presence reinforces the project's importance and their personal investment.

- Empowering the right internal champions and business owners: While consultants bring methodology and technical know-how, the ultimate responsibility for business process design and adoption must rest with internal leaders. Executives empower key business leaders to serve as process owners for critical areas (e.g., "order-to-cash process owner," "procure-to-pay process owner"). These individuals are accountable for defining the future state of their processes, driving user adoption, and ensuring the SAP solution truly meets business needs. They are the bridge between the strategic vision and operational reality.

- Establishing robust governance structures: Leaders establish and actively participate in governance bodies (e.g., a Project Steering Committee, a Design Authority Board). These committees are not just rubber stamps; they are forums for resolving conflicts, making critical decisions, approving scope changes based on the vision, and holding the project team accountable for progress against objectives. They ensure that all decisions consistently point back to the agreed-upon North Star.

- Guardrails for conflict resolution: When interdepartmental conflicts or disagreements over design choices inevitably arise, leaders provide immediate clarity and decisive resolution. Instead of leaving these complex decisions to consultants (who often lack the authority or context to truly arbitrate), executives step in, pointing back to the overarching vision and nonnegotiable principles to guide the path forward.

When this level of leadership ownership is present, external consultants still play an absolutely vital and indispensable role—but their role shifts from being the drivers of the vision to being expert enablers and trusted advisors who help bring your vision to life. They provide the "how" to your "why."

THE CONSULTANT'S ESSENTIAL ROLE IN VISION: CHALLENGER AND ENABLER

If consultants cannot own the vision, then what is their proper and incredibly valuable role? Their job is to be the expert challenger and the indispensable enabler. The challenger is a good, ethical consultant who will push back if they encounter a vague, inconsistent, or unrealistic vision. They will ask incisive, tough questions designed to provoke clarity: "What specific business problem are you trying to solve here?" "How does this proposed solution align with your stated strategic objectives?" "Have you considered the long-term implications of this design choice?" They leverage their cross-industry experience and SAP best-practice knowledge to help the client refine and sharpen their own vision, ensuring it is grounded in reality and maximizes the potential of the SAP platform. They act as a critical sounding board, but they do not dictate.

Once the vision is clearly defined, owned, and articulated by the client's leadership, the consultants shift into their role as powerful enablers. They translate that strategic vision into a concrete design, execute the complex configuration, manage the technical development, and guide the deployment. They bring the methodological rigor, the technical expertise, the project management discipline, and the sheer human resources required to bring the client's vision to life within the SAP system.

At no point should consultants dictate the strategic "why" or "what" for your business. Their expertise lies in the "how"—the execution, optimization, and technical delivery—within the framework of your defined direction. It's a true partnership, with clear lines of responsibility.

THE POWER OF VISION OWNERSHIP

To underscore the profound impact of vision ownership, let's examine two contrasting, yet very real, examples drawn from countless SAP projects I've observed.

Company A: The Outsourced Vision—a Cautionary Tale

Company A, a large, established manufacturing firm, was facing an aging legacy ERP system that was becoming increasingly costly to maintain and hindering their ability to scale. The executive leadership, busy with day-to-day operations, handed almost complete control of their SAP S/4HANA implementation to a prestigious, well-known consulting firm. The implicit assumption was, "They're the experts. They've done this before, so they know best."

In the absence of a strong, internally owned vision, the project quickly devolved. Every department's request, no matter how minor or how much it deviated from standard SAP functionality, was treated with equal importance. Without a clear "no" criterion from the top, the consultants, eager to please their client and avoid internal friction, acquiesced to nearly every customization request. Design workshops became battlegrounds where departments fought for their unique, often legacy-bound, processes to be replicated in SAP.

By the time of go-live, the system, while technically functional, bore an uncanny resemblance to their old legacy ERP, albeit with a modern interface. It was heavily customized, cumbersome, and incredibly complex. Employees, having rarely been engaged in defining the "why" or the "what" beyond their immediate tasks, greeted the new system with a collective shrug: "What was the point of all this massive effort and expense?" User adoption lagged significantly, the promised efficiency gains were negligible, and the total cost of ownership ballooned due to the extensive custom code and ongoing support needs. Within two years, Company A was already contemplating replacing SAP with another solution, effectively wasting tens of millions and years of effort.

Company B: The Owned Vision—a Story of Transformation

Company B, a rapidly expanding global services provider, recognized that their disparate systems were hindering their international growth strategy. Their executive leadership, led by a highly engaged CEO, made an up-front, nonnegotiable commitment: Their SAP project's singular, overarching goal was to create one global template for finance, human resources, and core operations. Their vision was "global, integrated, and standardized for agility."

During the design phase, when individual country divisions or specific departments inevitably pushed for exceptions or custom functionalities, the executive team didn't defer to the consultants. Instead, they would directly challenge the requestors with a simple, powerful question: "Does this deviation align with our vision of a global, standardized template? If not, what is the overwhelming strategic imperative that justifies the added cost, complexity, and risk?" If a compelling, undeniable business or legal reason couldn't be articulated, the answer, frequently, was a firm but polite no.

This unwavering commitment to the owned vision empowered the project team and consultants to push back effectively against unnecessary customization. The project finished on time and largely within budget. User adoption was high because employees understood the strategic imperative and saw leadership modeling the desired behavior. Within months of go-live, Company B was able to seamlessly integrate newly acquired entities onto the global template, expand into new markets with unprecedented speed, and gain real-time financial insights that were previously unimaginable. The SAP system became not just a new piece of software but the foundational platform for their accelerated global expansion.

The stark difference between these two companies was not the sophistication of the technology, the caliber of the consultants, or the complexity of their businesses. It was, unequivocally, the degree of leadership ownership of the strategic vision.

DEFINE AND PROTECT YOUR VISION: ACTIONABLE STEPS FOR EXECUTIVES

Owning and protecting the SAP vision is easier said than done, particularly amid the daily pressures of running a business. However, it is an achievable and critical endeavor. The practical and actionable steps executives can take start with asking "why," and keep asking it. Before any technical workshops, before any vendor pitches, the executive team must dedicate focused time—perhaps an off-site workshop—to definitively agree on the core business drivers for the SAP investment. Articulate these drivers in plain, unambiguous language. Write them down. Share them widely and consistently. This "why" should be the first slide in every project presentation and the central theme of every major communication.

If the executive team (CEO, CFO, CIO, COO, CPO, etc.) is not in full, genuine alignment on the SAP vision, the organization below them will inevitably become fragmented and pull in different directions. Dedicate time to resolve internal differences, conflicting priorities, and historical biases before the project truly begins. This might involve difficult conversations, but resolving them up front is far less costly than managing them during the implementation. Here are actionable steps to protect your SAP vision:

- Translate vision into actionable principles and guardrails: Move beyond abstract statements. Transform your vision into concrete, actionable design principles that guide daily decision-making for the project team. Examples include:

  - "Prioritize standard SAP functionality over customization wherever possible."

  - "Any customization must demonstrate a clear, quantifiable ROI and competitive advantage."

  - "Data quality is a shared business responsibility, not an IT task."

  - "Simplify existing processes to align with SAP best practices rather than replicating complexity." Embed these principles into project charters, design documents, and decision-making frameworks.

- Revisit and reinforce constantly: An SAP project spans months, often years. Vision fatigue is real. Leaders must proactively combat this by repeating, reinforcing, and reenergizing the vision constantly. This means

- Regular project updates from the executive sponsor

- "Town hall" meetings where leaders connect project progress to the strategic vision

- Visual reminders (e.g., posters, intranet banners) of the North Star

- Celebrating milestones and explicitly linking them back to the vision

- Empower the right internal champions (business leaders, not just IT): Appoint highly respected, knowledgeable business leaders—not just IT personnel or external consultants—as ultimate process owners and decision-makers for their respective functional areas. These champions must understand the vision, be empowered to make decisions that align with it, and be accountable for driving adoption within their teams. Consultants advise; business leaders decide and own.

- Establish a clear governance model: Formalize how decisions will be made, how conflicts will be resolved, and how scope changes will be managed. The governance model (e.g., a steering committee, a design authority board) should always have the vision at the core of its decision-making criteria. This prevents ad hoc, politically driven choices.

These steps, when consistently applied, create a robust internal shield that prevents the vision from being diluted, compromised, or outright lost as the project unfolds amid its inherent complexities.

When companies commit to the discipline of defining and passionately owning their SAP vision, the long-term payoffs are substantial and transformative, extending far beyond the project's go-live date. With a clear North Star, endless internal debates become a thing of the past. Every project decision, no matter how minor, can be quickly assessed against the shared vision. This accelerates progress, reduces rework, and dramatically improves efficiency throughout the implementation.

When the vision is clear, consistent, and championed by internal leaders, employees understand that the SAP project is not merely "the consultant's project" or "an IT mandate." It becomes "our company's transformation," a shared journey toward a better future. This sense of collective ownership fosters engagement, reduces resistance, and accelerates user adoption, ensuring the new system is fully leveraged.

An SAP system built upon a strong, owned vision becomes a dynamic platform, not a static piece of software. It is inherently more flexible, scalable, and easier to upgrade. This enables the organization to adapt to market changes, embrace new business models, and integrate future acquisitions seamlessly, ensuring the SAP investment continues to deliver value and competitive advantage for decades.

By focusing on the strategic "why," organizations are far more likely to achieve the tangible business benefits outlined in their original business case—from cost savings and efficiency gains to improved customer satisfaction and enhanced decision-making capabilities. The vision ensures the project delivers real business value, not just technical functionality.

The rigorous process of defining and owning the SAP vision builds invaluable internal capabilities. It strengthens cross-functional collaboration, enhances strategic thinking, and develops the muscle for managing large-scale organizational change. These newly acquired skills are transferable, preparing the organization for future digital transformation initiatives.

This is the fundamental difference between SAP projects that deliver deep, long-term transformation and those that, despite massive investment, ultimately fade into irrelevance or become perpetual liabilities.

The lesson regarding vision in SAP implementation is remarkably simple, yet its implications are profound: Do not outsource your vision. Your

SAP project will ultimately succeed, or tragically fail, not because of the inherent capabilities of the software, nor solely due to the methodology employed, but predominantly because of the clarity, conviction, and consistent leadership you bring to its strategic direction.

If you are an executive reading this, internalize this fundamental truth: The vision for your SAP transformation must emanate from you and your fellow senior leaders. This is not because you are expected to possess the most intricate technical knowledge of SAP—that's the role of your consultants and internal IT experts. It is because you, and only you, possess the deepest, most nuanced understanding of your business: its unique competitive landscape, its strategic objectives, its market position, its cultural DNA, and its long-term aspirations. This unparalleled knowledge of your business cannot be bought, borrowed, or delegated. It must be cultivated, owned, protected, and lived every single day of the project's demanding life cycle.

The consultants you engage are indispensable partners. They will provide the expertise, the framework, and the horsepower to execute. But you, the leader, must provide the destination, the map, and the compass. That unwavering direction is the most powerful asset you bring to the table.

Once your vision is firmly established and wholly owned, the critical subsequent question becomes this: How do you translate that aspirational vision into a concrete, actionable strategy that will stand the test of time, deliver measurable results, and truly transform your enterprise?

Why Most SAP Projects Fail Before They Begin

It's a sobering, yet frequently observed truth in the world of enterprise technology: Most SAP projects don't actually fail at the highly visible go-live moment, or even during the rigorous testing phases. Instead, the seeds of their demise are sown much, much earlier—sometimes even before the first external consultant sets foot in your office, or before the formal project kickoff meeting is scheduled. The fundamental cracks in the foundation, the critical misalignments, and the crucial decisions made for the wrong reasons occur in the earliest days. By the time the "real project" involving detailed configuration, development, and extensive testing begins, the damage is already done, often irrevocably.

Understanding these insidious pitfalls is the first step toward building a truly resilient and successful SAP transformation. Executives, eager to demonstrate progress and fulfill mandates, often operate under a dangerous illusion of readiness. They genuinely believe their organization is prepared for an SAP implementation simply because they've gone through the motions: A budget has been approved, a multimillion-dollar contract has been signed with a system integrator, and a steering committee, often comprising senior leaders, has been assigned. While these are necessary procedural steps, they are, in isolation, woefully insufficient.

True readiness transcends mere paperwork and financial allocations. It is fundamentally about a state of profound alignment, crystal clear clarity, and unflinching organizational honesty. It requires a deep, shared understanding of the undertaking and an unwavering commitment to its strategic imperatives.

True readiness compels leaders to honestly answer critical, often uncomfortable questions long before any software is installed or configured. Have we, as an executive team, unequivocally defined why we are undertaking this monumental SAP project now? Is it a strategic imperative for growth, a path to market dominance, or a fundamental shift in our operating model? Or is it merely a reactive response to an aging system or competitor actions?

Can we articulate the quantifiable business benefits (e.g., reduced inventory costs, faster financial close, improved customer satisfaction) that will validate this massive investment? Do we possess a shared, tangible vision of what "success" truly looks like for our business beyond simply "going live"?

This isn't about just identifying problems but committing to addressing them before they become showstoppers. Is our master data clean enough? Are our processes documented? Have we conducted an honest, rigorous assessment of whether our existing data, current business processes, and, most critically, our people are genuinely ready for a transformation of this magnitude? Is our culture prepared for significant change?

In an alarming number of projects I've witnessed falter, the candid answer to these foundational questions was a resounding "not really." Leaders often harbored an optimistic, yet misguided, belief that they could "figure it out along the way." They assumed that the highly paid consultants would somehow "sort it all out" once they arrived. This passive approach is fatal. SAP, as a highly integrated and disciplined system, is notoriously unforgiving of a weak, unprepared, or dishonest foundation. The technical brilliance of the software cannot compensate for fundamental strategic, operational, or cultural deficiencies.

COMMON EARLY MISTAKES THAT DOOM PROJECTS: CRACKS IN THE CORNERSTONE

The early days of an SAP project are a critical window where foundational mistakes are made, often innocently, but with devastating long-term consequences. These errors, if unaddressed, create cracks in the cornerstone, leading to instability and eventual collapse.

Mistake 1: Treating SAP Like a Mere Technology Upgrade

This is arguably the most prevalent and destructive mindset that poisons projects from their inception. Too many organizations approach SAP with a simplistic view: "Our current system is old and inefficient. SAP is new and modern; therefore, let's simply swap it out." This perspective reduces a profound business transformation into a mere technical refresh, akin to upgrading from an old car to a newer model of the same make.

The critical fallacy here is that SAP is not merely new software; it is a new operating model for your entire business. It encapsulates industry best practices and demands a disciplined, integrated way of working. If leaders treat SAP implementation solely as an IT project—focused on server specs, network latency, and software versions—they fundamentally miss the enormous opportunity for process optimization, data unification, and strategic agility. They fail to challenge outdated business processes, continue to operate in departmental silos, and miss the chance to reinvent their organization. Such projects may "go live" technically, but they fail to deliver genuine business value, becoming an expensive technical replacement rather than a strategic enabler. The project fails before kickoff because the ambition was too low and the understanding of SAP's true power was too shallow.

Mistake 2: Chasing "Best Practices" Without Context or Critical Thought

Consultants, with good intentions, often evangelize "best practices" embedded within SAP or derived from their cross-industry experience. This often leads to the dangerous notion that adopting SAP means blindly adopting every single one of these so-called best practices. However, what constitutes a "best practice" for a highly specialized pharmaceutical manufacturer in Germany might be entirely inappropriate, or even detrimental, for a fast-moving consumer goods distributor in Chicago or a bespoke services firm in London.

Projects falter when companies blindly attempt to "lift and shift" best practices without the following:

- Rigorous contextual analysis: Understanding why a particular process is a "best practice" and how it aligns with your unique strategic goals, competitive differentiators, and regulatory environment

- Critical internal assessment: Determining which existing processes are truly inefficient and need to be replaced by a standard best practice versus those that provide a genuine competitive advantage and might warrant careful consideration for adaptation (though this is rare)

- Adaptation, not blind adoption: Recognizing that even best practices often require careful adaptation to an organization's specific culture, scale, and market dynamics

A project that blindly chases "best practices" without critical thought risks imposing an ill-fitting operational model on the business, leading to internal resistance, frustrated users, and a system that fails to support the company's unique value proposition. It effectively allows external templates to dictate internal strategy.

Mistake 3: Underestimating the "Work Before the Work"

The allure of diving straight into configuring the new system is powerful. Executives and project teams alike want to see tangible progress. However, many companies catastrophically underestimate—or simply skip—the vital "work before the work." This includes the following foundational activities that must happen before deep design or configuration workshops can be productive:

- Comprehensive data assessment and cleansing: Data is the lifeblood. Failing to conduct a thorough health check of master data (customer, vendor, material, financial data) and initiating significant cleansing efforts before the project gets into full swing are a recipe for disaster. Migrating "garbage" into SAP only digitizes and amplifies chaos.

- Process harmonization and simplification: Before designing processes in SAP, businesses need to map their current processes, identify redundancies, eliminate non-value-added steps, and agree on harmonized "to be" processes across departments. This isn't about fitting SAP to current processes; it's about redesigning processes for SAP.

- Early change impact analysis: Understand who will be affected by the new system and how their jobs will change, long before formal training begins. This allows for proactive communication and targeted engagement strategies.

- Establishing data governance: Define clear ownership and account-ability for master data before the system is designed, rather than figuring it out after go-live.

Rushing into configuration workshops without this foundational preparation leads to the project becoming perpetually bogged down in firefighting. Decisions are made reactively, rework becomes endemic, and initial momentum rapidly dies, causing delays and budget overruns.

Mistake 4: Delegating Instead of Leading

As explored extensively in Chapter 2, the executive leadership team is the ultimate steward of the SAP vision. Yet an alarmingly common early mistake is the tendency to delegate this core leadership responsibility to consultants, believing they can "drive the bus." Leaders retreat into an oversight role, assuming their job is merely to approve budgets and attend steering committee meetings, leaving the strategic heavy lifting to external experts.

Without sustained, visible executive leadership and ownership of the vision, the project lacks internal authority and direction. Conflicts between departments (Finance vs. Sales, Operations vs. Marketing) remain unresolved because no one internal has the mandate or the strategic clarity to make the tough calls. Consultants, despite their expertise, cannot force internal alignment or make strategic trade-offs that only the business owners can. This delegation of leadership transforms the project from a strategic business initiative into a technical implementation, often leading the "bus" straight into a ditch of misaligned goals, uncontrolled scope, and internal resistance.

Mistake 5: Ignoring the Culture Factor and Human Element

An SAP implementation is, at its heart, a massive organizational change initiative. It touches every process, every role, and every employee. Yet many projects fail before they begin because leaders fundamentally ignore, or significantly underestimate, the pervasive culture factor and the critical human element.

Ignoring culture can manifest in several ways:

- Lack of early communication: Employees are kept in the dark until late in the project, leading to rumors, fear, and resentment.

- Failure to address resistance: Not proactively identifying and addressing the natural human resistance to change. Employees fear job loss, loss of control, or simply the unknown.

- Insufficient stakeholder engagement: Not involving key business users and process owners early in the design phase, leading to a feeling of "this is being done *to* us, not *with* us."

- Underestimating training needs: Assuming a few days of training at the end will suffice for users to become proficient in a completely new way of working.

If leaders haven't actively prepared the organizational culture for the impending transformation—fostering open communication, building trust, and creating a compelling "what's in it for me" narrative for employees—then even the most technically elegant and well-designed system will be sabotaged by passive resistance, low adoption, and, ultimately, a failure to deliver the intended benefits. The project can be technically perfect but organizationally rejected.

THE FIRST NINETY DAYS DECIDE EVERYTHING

The initial three months of an SAP project are often deceptively quiet. There's a flurry of activity—kickoff meetings, discovery sessions, endless documentation reviews, and preliminary workshops. On the surface, it might feel like "nothing major" or truly transformative is happening. This quiet period, however, is precisely the most dangerous and determinative time for the entire project's trajectory.

The work done (or neglected) in these first ninety days forms the bedrock upon which every subsequent decision, design choice, and configuration will be built. If the early work is not executed with absolute rigor and clarity, every decision made later is inherently built on shaky ground:

- Wrong assumptions go unchallenged: If the initial understanding of current processes, data quality, or business priorities is flawed, these fundamental errors propagate throughout the design and build phases, creating systemic issues that are incredibly difficult and costly to fix later.

- Misaligned goals become embedded as design principles: When the executive team lacks alignment, conflicting departmental objectives can inadvertently become core components of the system design, creating internal process friction that persists after go-live.

- Data gaps become critical showstoppers: If data cleansing and governance aren't prioritized, the project will invariably hit a wall during migration or testing, leading to significant delays and budget overruns as emergency data remediation efforts are launched.

- Cultural resistance solidifies: If employees are left out of the loop or feel their concerns are ignored during the early stages, their initial skepticism can harden into active resistance, making later change-management efforts exponentially more difficult.

I often use an analogy with clients: "You don't pay for mistakes precisely when you make them in an SAP project; you pay for them six, nine, or twelve months later, with substantial interest." A flawed decision in the first ninety days can cost ten times more to fix at go-live than it would have in the initial discovery phase. This exponential cost is why the strategic focus on the early innings is absolutely nonnegotiable.

A TALE OF TWO KICKOFFS: INVESTING IN CLARITY VS. RUSHING TO START

To highlight the profound impact of these early choices, let's examine two real-world examples (names and industries disguised to protect confidentiality, but the scenarios are drawn directly from my experience).

Company A: The Rush to Start—a Predictable Downfall

Company A, a diversified industrial conglomerate, was under immense pressure from its board to "modernize its IT infrastructure" and initiate an SAP S/4HANA project to replace an aging, fragmented ERP landscape. The CEO, keen to announce progress at an upcoming board meeting, pushed for an accelerated project kickoff.

The result? The kickoff event was held before any significant internal assessment of existing data quality, detailed business process mapping, or deep alignment on strategic business drivers. Initial design workshops began almost immediately, but with no overarching, clearly defined business priorities. Different departmental leaders (Finance, Operations, Sales) came to the table with their own siloed agendas and legacy requirements.

Six months into the project, the entire initiative ground to a near halt. Why? Finance adamantly demanded one set of processes and controls, Operations insisted on another for their specific production methods, and Sales required a third for their unique customer engagement model. With no clear, unifying vision from the executive sponsor (who had largely disengaged after kickoff), and consultants who lacked the internal authority to arbitrate, no one could agree on a standardized path forward. The project spiraled, scope ballooned uncontrollably, costs escalated dramatically, and internal trust among stakeholders collapsed. The project struggled for years before a highly pared-down, compromised system eventually went live, delivering a fraction of the promised value.

Company B: The Careful Beginning— the Power of Foresight

Company B, a global consumer goods company, also recognized the urgent need for a new ERP system to support its rapid international expansion. However, their leadership took a fundamentally different approach. Before even scheduling a formal project kickoff, the executive team, led by an engaged CIO and COO, dedicated a full six weeks to intensive internal alignment workshops. Their focus was singular: to define precise, nonnegotiable business priorities for the SAP transformation. These included "achieving a faster financial close cycle," "harmonizing global product data for consistent analytics," and "establishing a single, streamlined order-to-cash process across all international entities."

Crucially, they also invested heavily in an extensive master data assessment and began initial data cleansing efforts before the system integrator even commenced work. The formal project kickoff, therefore, occurred later than Company A's, but when it finally happened, everyone involved—from the executive steering committee to the core project team and key business users—was remarkably aligned on the "why" and the "what."

As the project progressed, conflicts inevitably arose, but they were quickly resolved by referencing the clearly articulated business priorities. The consultants knew the boundaries and could effectively push back on unwarranted customization. Company B's project moved significantly faster, not slower, than expected, precisely because they had invested heavily in clarity and preparation up front. The outcome was a highly successful go-live, robust user adoption, and the ability to expand into several new markets seamlessly within months, directly attributable to the standardized and integrated SAP platform.

The stark difference between these two outcomes was not the choice of SAP version, the size of the consulting firm, or the underlying technology; it was the discipline, courage, and commitment to strategic clarity demonstrated in the very first ninety days of the project.

THE HIDDEN ENEMIES OF EARLY SUCCESS: PRESSURES DISGUISED AS LOGIC

Projects don't fail because leaders are inherently lazy or intentionally negligent. Often, they falter because leaders are pressured by forces that, on the surface, appear logical and compelling but are, in reality, deeply destructive to the foundational work required for success. These "hidden enemies" create a false sense of urgency and push companies into premature starts.

One such enemy is the calendar trap: "We need to start now to meet next year's critical go-live date/reporting deadline/peak season." This rigid adherence to an arbitrary timeline, often set without proper understanding of the foundational work, forces a superficial readiness. It creates immense pressure to skip critical assessment and alignment phases, leading to corners being cut that will prove exponentially costly later.

Another is the budget trap: "We've allocated significant funds for this project in the current fiscal year, so we must begin immediately to utilize the budget and show progress." This financial imperative often overrides strategic common sense. It incentivizes activity over meaningful preparation, leading to rushed decisions and a rapid burn of funds on a weak foundation.

There is also the competitive trap: "Our closest competitor just announced their SAP transformation, so we must do the same, and faster, to maintain our market position." This fear-driven reaction to competitive announcements often leads to a "keeping up with the Joneses" mentality, pushing companies to start without truly understanding their unique business drivers or preparing their internal organization. The focus shifts from internal value creation to external perception.

Then there is the "legacy pain" trap: "Our current systems are so bad; we simply cannot endure them any longer. We need to start ASAP to alleviate the pain." While legacy system pain is a legitimate driver, it can lead to a desperate rush that overlooks the necessary groundwork. Jumping from one painful system to another, without fixing the underlying issues, often just exchanges one set of problems for another, more expensive set.

Finally, there is the "fear of opportunity cost" trap: Leaders fear that by delaying, they are missing out on potential benefits or competitive advantage. This can be true if delays are due to indecision, but strategic patience—investing time up front to get it right—ultimately delivers greater returns and avoids the far higher opportunity cost of a failed or compromised project.

These powerful drivers push companies to initiate SAP projects before they are genuinely ready, before the strategic "why" is truly defined, and before the internal organization is aligned. And once the project gains momentum, with consultants on the payroll and internal teams formed, it becomes almost psychologically and financially impossible to hit pause—even when the brewing problems are becoming glaringly obvious. The inertia of a live project can be incredibly difficult to overcome.

HOW TO AVOID FAILING BEFORE YOU BEGIN

The good news is that early failure is predictable, and therefore preventable. It requires courage, discipline, and a willingness to challenge conventional project wisdom. Here are the nonnegotiable disciplines that forge strong beginnings:

- Define the business case beyond IT (the "why" beyond the "what"): Ensure your business case for SAP is rooted in quantifiable business outcomes: growth enablement, significant efficiency gains, enhanced customer experience, or a fundamental shift in your operating model. It must be a compelling narrative that goes far beyond simply "replacing old technology." This strategic clarity is the project's North Star.

- Align the executive team first (lock them in a room if necessary): This cannot be overstated. Before any significant funds are spent on external consultants, the executive leadership team must achieve absolute, unequivocal alignment on the vision, strategic priorities, and nonnegotiable design principles for the SAP project. Resolve all disagreements and conflicting departmental agendas up front. A united leadership team is the single greatest predictor of project success.

- Do a comprehensive data health check and start cleansing early: Before a single line of configuration, invest in a thorough assessment of your master data quality (customers, vendors, materials, financial accounts). If it's messy, commit resources to begin cleansing and structuring it immediately. Data migration is a project within a project, and it can become a showstopper if not proactively managed from day one.

- Simplify the scope (relentlessly): Complexity is the number one killer of SAP projects. Resist the urge to include every "nice to have" feature or to replicate every legacy process. Rigorously define what is essential (minimum viable product) versus what is optional for the initial go-live. A phased approach that delivers core value quickly is almost always preferable to an overly ambitious "big bang" that attempts to do everything at once. Be prepared to say no to nonstrategic demands.

- Plan the change journey early (communicate and involve): Begin communicating to employees that an SAP transformation is coming, why it's happening, and what it means for them well in advance of the technical implementation. Involve key business users and process owners in the design phase, making them cocreators of the solution, not just passive recipients of a new system. This proactive change management mitigates resistance and builds early buy-in.

- Resist false urgency (embrace strategic patience): Do not let artificial deadlines, budget cycles, or competitive pressures force a premature start. Strategic patience is a virtue in SAP. Starting one month later with absolute clarity, executive alignment, clean data, and engaged people is exponentially better—and ultimately faster—than starting today in chaos, indecision, and a state of unreadiness.

THE CONSULTANT'S WARNING SIGNS: SIGNALS FROM THE FRONT LINES

Experienced consultants, working closely with your team, often develop a keen sense for whether a project is heading for trouble, sometimes within the first month. As a leader, it's crucial to recognize their implicit, or sometimes explicit, warning signs. If you hear phrases like these, or observe these behaviors, your project may already be offtrack, and it's time to pause and intervene:

- "We'll figure that out later" (regarding a critical, foundational decision): This is a red flag, especially if it pertains to vision, core process design, or data. Deferring fundamental decisions means building on unstable ground.

- "Let's capture all requirements first" (without prioritization or fit-to-standard discussions): This indicates a lack of scope discipline and a tendency toward "lift and shift" rather than true transformation. It's a precursor to over-customization.

- "We don't have time to align on that right now" (when discussing executive or cross-functional alignment): This signals a fundamental lack of leadership commitment to the essential "work before the work." It means critical strategic conflicts are being ignored rather than resolved.

- "The business just needs to decide what they want" (without proper facilitation or data): This indicates consultants feel a lack of internal direction and are being asked to solve internal political or strategic challenges.

- "We can build that custom" (as a default solution to a gap): While sometimes necessary, if this is the immediate response to every deviation from standard, it indicates a project team or consultant leaning too heavily on customization rather than process reengineering.

- Lack of business engagement in early workshops: If business leaders and key users are consistently absent or disengaged in foundational workshops, it means critical perspective is missing, and the solution will likely not meet their needs.

These are not merely operational issues; they are deep signals that the project's foundation is weak, its strategic direction is unclear, or its leadership is disengaged. Smart leaders recognize these warnings and have the courage to pause, conduct a rapid diagnostic, realign, and then—only then—continue the journey. Ignoring them is to accelerate toward a predictable failure.

The truth is this: The failure of an SAP project is rarely a mysterious, unforeseeable event. It is, almost without exception, predictable. If a company begins such a complex transformation without a clear, owned vision, without robust executive alignment, without thorough data preparation, and without proactively engaging its people, then failure—in some form, whether outright collapse or a crippled, underperforming system—is almost certain. The path of least resistance often leads to the greatest cost and disappointment.

Conversely, companies that exhibit the discipline to slow down in the early stages, meticulously define their "why," ensure genuine executive and organizational alignment, invest heavily in data readiness, and proactively prepare their culture for change almost always succeed in delivering truly transformative value from their SAP investment.

The difference between these two outcomes is not a matter of luck, superior software, or even the size of the budget; it is, fundamentally, a matter of discipline, courage, and strategic foresight in the critical, often quiet, beginning.

The profound lesson is simple yet critical: Most SAP projects don't collapse at the highly public stages of testing or go-live; they collapse at kickoff. They fail because the foundation was never laid properly or because the early cracks were ignored. And once you're six, nine, or twelve months into a multimillion-dollar project, no amount of heroic effort, additional budget, or brilliant technical fixes can fully undo those fundamental early mistakes. The compounding effect of poor early decisions is devastating.

The message to every leader contemplating or embarking on an SAP project is therefore clear and urgent: the beginning truly decides the end. Invest the necessary time, commit the significant effort, and muster the courage to get the foundational work absolutely right up front. Define the compelling "why" for your organization. Achieve genuine, nonnegotiable alignment among your executive team. Ruthlessly clean and prepare your data. Proactively simplify your scope. And, crucially, prepare and engage your people for the immense change ahead.

Because if you don't, your multimillion-dollar SAP project may already be failing—before it even begins.

The Software Illusion: Strategy over Software

When executives put their signature on the dotted line for a multimillion-dollar SAP implementation, a powerful wave of excitement often sweeps through the organization. The promise of a shiny new system—a cutting-edge enterprise resource planning (ERP) platform—is intoxicating. It pledges to finally replace decades of cumbersome patchwork processes, unreliable manual spreadsheets, and the pervasive frustrations of outdated legacy systems. SAP, in this euphoric vision, becomes more than just software; it transforms into the tangible symbol of modernization, the rallying banner under which leaders confidently declare to their people, "This is our future! This is how we will achieve unprecedented efficiency, global reach, and competitive dominance!"

But here lies the fundamental and often devastating reality: Software alone does not transform a business. No matter how technologically powerful, intricately configurable, or supposedly "intelligent" an ERP system like SAP may be, it is inherently a tool. And like any tool, its effectiveness is entirely dependent on the clarity—or disconcerting lack thereof—of the strategy that guides its application. Without a precise and well-defined purpose, even the most sophisticated instrument can be wielded ineffectively, leading to disappointing, or even detrimental, outcomes.

Too many SAP projects, unfortunately, stumble and ultimately fail because leaders unconsciously harbor a dangerous belief: that merely acquiring SAP, by signing a contract and initiating the project, is synonymous with purchasing business transformation itself. They expect the software, through its inherent capabilities and "best practices," to miraculously "save" them from operational inefficiencies, data fragmentation, and competitive pressures. The truth, however, is significantly harsher and far more critical: SAP is not the strategy; it is, and can only ever be, a powerful enabler of strategy. Without a crystal clear business vision and a meticulously defined operational plan to achieve that vision, SAP will merely digitize existing confusion, automate deeply ingrained inefficiency, and scale inherent dysfunction across the enterprise. It becomes a very expensive mirror reflecting your existing organizational chaos rather than a powerful engine for improvement.

The profound role of SAP is not to dictate to a company who it should be, what its market position should be, or what its core competencies ought to be. Its purpose is precisely the opposite: to embody, operationalize, and provide the technical backbone for a company's existing strategic intent. If leadership lacks a clear understanding of where the business is heading, what markets it aims to conquer, or what customer experience it seeks to deliver, no amount of advanced software can possibly fill that fundamental strategic void.

Consider two hypothetical companies, both prominent global manufacturers, each investing tens of millions of dollars—and countless hours of human capital—into a comprehensive SAP implementation.

Company X approaches its SAP project primarily as an IT project. The focus is on technical migration, system uptime, and feature parity with their old system. Business leaders are involved, but their input is seen as "requirements gathering" for the IT solution rather than strategic decision-making. After go-live, the company is still embroiled in endless internal arguments about how to accurately calculate the cost of goods sold, because different departments (Finance, Operations, Sales) had conflicting views, and the SAP design simply accommodated all of them, leading to convoluted processes and unreliable data. The software works, but the business processes remain fragmented and inefficient, never truly leveraging SAP's integrated capabilities.

Company Y approaches its SAP project as a profound business reinvention project. From the outset, the executive leadership team defines a clear strategic objective: "Become the most agile and cost-effective manufacturer in our segment by optimizing our global supply chain and empowering data-driven financial decisions to improve overall margin." Every SAP design choice—from master data structure to inventory management processes and financial reporting hierarchies—is rigorously measured against this single, unifying margin improvement strategy. After go-live, Finance, Operations, and Supply Chain leaders are seamlessly aligned around this strategy, and SAP becomes the indisputable backbone of their execution, providing real-time insights into costs, identifying margin leakage, and enabling swift corrective actions.

The two companies invested in the same core SAP software. They had access to the same powerful features and functionalities. Yet they achieved two profoundly different outcomes. The defining factor was not in the lines of code or the server specifications; it was, unequivocally, in the clarity, coherence, and executive ownership of their guiding business strategy. SAP simply amplified what was already present—or absent—in their strategic foundation.

One particular client that I worked with was a thriving mid-market manufacturer specializing in consumer durables. They recognized the need to replace its aging, highly customized legacy ERP system. Their approach to the SAP implementation was classic "technology upgrade." The CIO was designated the primary project sponsor, and the directive handed down from the CEO was succinct and seemingly straightforward: "Get us live on SAP in eighteen months. We need a modern system."

What was conspicuously absent was any deep, sustained engagement with broader business leadership (CFO, COO, head of Sales, etc.) to define a strategic vision beyond just "modernizing." There were no dedicated strategy sessions exploring how SAP could enable future growth, fundamentally enhance customer experience, or provide a sustainable competitive differentiation. The project became an internal IT initiative, driven by a tight timeline.

The predictable result? The project team, comprising internal IT staff and external consultants, spent endless, agonizing cycles debating minute configuration options without any clear business North Star to guide their decisions. Should pricing logic follow complex customer contracts or be simplified around product families? Should manufacturing planning be managed regionally with local autonomy, or centralized for global optimization? How should new product introductions be handled from a data perspective? Without strategic principles to serve as tiebreakers, the SAP system design meetings devolved into a battlefield of competing departmental opinions, political maneuvering, and historical biases. Each department fought for its existing processes to be replicated, leading to an overly complex, heavily customized solution.

Go-live eventually happened—after significant delays and substantial cost overruns. But the system was perceived as cumbersome. User adoption was weak because the new processes felt arbitrary and did not clearly align with perceived business objectives. Operational costs skyrocketed due to the complexity and ongoing support of the highly customized system. Within two years, the leadership, disheartened by the lack of tangible ROI and ongoing operational friction, seriously began to consider replacing SAP altogether, viewing it as "too complex" or "not fitting our business"—when, in reality, it was their own lack of strategic leadership that had automated their chaos.

Now, contrast that experience with another client: a fast-growing global distributor of specialized industrial components. Their CEO, with foresight and unwavering conviction, approached SAP not as an IT project but as an integral part of a bold, enterprise-wide transformation. Their stated strategic objective was clear and powerful: "To become the most customer-responsive player in our industry, leveraging real-time data to delight clients and optimize our supply chain agility." This strategic lens became the absolute measure for every decision.

During every design choice, every process discussion, and every conflict resolution, the executive team and the empowered business process owners consistently asked a single, guiding question: "Does this design choice, this configuration, this process flow, genuinely improve our customer responsiveness?" When conflicts arose between departments—for instance, between Finance's desire for rigid billing cycles and Sales's need for flexible order adjustments—the answer was simple: Whichever option got them closer to customers, faster, with greater accuracy and less friction, won the day. If a proposed customization did not directly contribute to superior customer responsiveness, it was rejected.

The difference in project dynamics was night and day. Decisions were made swiftly and purposefully. The system was designed to be lean, standard, and focused on enabling the core strategic objective. The project finished largely on time and within budget. User adoption was remarkably high—not just because of excellent training but because employees understood why they were changing and saw how the new system directly enabled their shared strategic goal of customer responsiveness. The company rapidly expanded into new global markets, seamlessly integrating acquired entities, because their SAP platform was built on a flexible, standardized strategic foundation. One project automated chaos; the other project brilliantly embedded strategy.

SAP, by its very nature, forces an organization to make thousands of detailed decisions during its implementation: how to manage customer master data, how to structure financial reporting, how to model complex supply chains, how to design the sales order process, and so much more. Without a clear, overarching guiding strategy, these myriad decisions become arbitrary, often political, and highly susceptible to individual preferences or departmental silos. With a dominant strategy, however, every single one of these choices becomes aligned and deeply purposeful, directly contributing to the defined business objectives.

Let's explore how strategic intent permeates seemingly technical design decisions:

- Financial strategy: If your primary strategic goal is granular, country-level profitability reporting to optimize local operations, you will design your profit centers, cost centers, and financial dimensions within SAP very differently than if your overarching goal is comprehensive, real-time global P&L visibility and consolidated financial statements for corporate mergers and acquisitions analysis. SAP can be configured to support both, but you must decide the strategic financial objective first. Without this, you might build a system that can do neither effectively.

- Supply chain strategy: If your strategic aim is to achieve absolute cost efficiency and maximize economies of scale, you will configure your supply chain processes within SAP to optimize for large, bulk production runs, centralized inventory management, and low-cost transportation routes. Conversely, if your strategic aim is market-leading responsiveness and speed to serve unpredictable customer demands, you will optimize for flexible production schedules, decentralized or distributed inventory points, and agile logistics networks. SAP, remarkably, can support either strategic approach. But the system's design, its data flows, and its reporting capabilities will always directly mirror your chosen priorities. A lack of clarity here leads to a system that tries to do everything and excels at nothing.

- Sales and customer strategy: If your business competes primarily on price leadership, your SAP configuration for sales processes will likely emphasize discounting mechanisms, complex rebate processing, and rigorous cost tracking to ensure margin adherence. If, however, your strategic differentiator is an unparalleled customer experience, your SAP design will prioritize features like faster quoting processes, highly accurate product configuration tools (e.g., variant configuration), seamless customer self-service portals, and integrated customer service workflows that provide a 360-degree view of the client. Again, SAP is a versatile tool, but its configuration will follow your strategic intent.

- Human capital strategy: Is your strategic goal to empower employees with self-service capabilities and streamline HR operations globally, or to centralize highly controlled talent management processes? Your SAP SuccessFactors or HCM module configuration will differ significantly.

- Research & development (R&D) and product life cycle management (PLM) strategy: If your strategy is rapid innovation and time-to-market, your SAP PLM design will focus on accelerating design cycles, collaborative engineering, and efficient material master creation. If it's about rigorous quality control and compliance, the emphasis shifts to robust quality gates, audit trails, and document management.

The powerful takeaway here is profound: Every single configuration choice, every process decision, and every data model definition in an SAP implementation is, at its core, a strategic choice in disguise. These are not merely technical adjustments; they are direct reflections of your business priorities.

THE PITFALLS OF SOFTWARE-FIRST THINKING: DIGITIZING DYSFUNCTION

When companies fall victim to the "software illusion" and forget that strategy must dominate software, they invariably fall into the trap of software-first thinking. This misguided approach typically manifests in several highly damaging ways, leading to significant financial waste, operational friction, and cultural resentment:

Endless Customization: The Fragile Frankenstein System

When the new SAP system doesn't inherently "fit" the business, because the business leadership hasn't clearly defined what that "fit" should strategically look like, the default reaction is often to demand customization. Consultants, often under pressure to deliver a solution that pleases every stakeholder, are then asked to code around every disagreement, every legacy process, and every departmental nuance. This results in the creation of a "Frankenstein system"—a patchwork quilt of bespoke code, complex interfaces, and nonstandard processes.

The consequences are dire and exponential:

- Astronomical costs: Custom development is expensive to build, test, and document.

- Increased fragility and risk: Customizations often break during SAP upgrades, introduce unexpected bugs, and complicate integrations with other systems. Each custom piece adds a point of failure.

- High maintenance burden: Supporting a highly customized system requires specialized knowledge and significantly more effort than maintaining standard SAP.

- Loss of best practices: The very benefit of SAP's embedded best practices is eroded as unique, often inefficient, legacy processes are replicated.

- Vendor lock-in: You become beholden to the knowledge of the specific team or individuals who built the customizations, making it harder to switch support partners.

In essence, uncontrolled customization creates fragile, expensive systems that eventually collapse under their own weight, becoming an albatross around the organization's neck.

Wasted Spend and Missed Opportunity: The Empty Investment

Companies that operate with a software-first mentality often burn millions of dollars implementing SAP features, modules, or functionalities that they ultimately never use, or use only minimally. This happens because there was no rigorous strategic filter applied to what genuinely mattered to the business. The procurement of software licenses and implementation services might proceed on the assumption that "more features are better," but without a clear strategic purpose for each feature, the investment yields little to no return.

Beyond the direct wasted spend, there's also the significant opportunity cost. Resources—financial, human, and time—are diverted to a project that fails to deliver its transformative potential. This means the business continues to suffer from existing inefficiencies, misses out on competitive advantages, and delays its ability to innovate or expand. The initial "investment" becomes a sunk cost, with very little to show for it in terms of improved business performance.

Cultural Frustration and Disengagement: The Human Cost

Perhaps the most insidious pitfall of software-first thinking is its profound impact on the organizational culture and employee morale. When SAP is implemented without a clear "why" and without linking the changes to a compelling business strategy, employees see the new system not as an enabler or a tool for shared goals but as an arbitrary burden. This disconnect often manifests in several predictable and damaging ways:

- "Extra work": They perceive the new processes as making their jobs harder, forcing them into workflows that make little strategic sense, or adding unnecessary complexity.

- Lack of ownership: They view it as "the consultants' project" or "IT's project," not "our company's transformation."

- Resistance and cynicism: Morale drops, user adoption suffers, and active or passive resistance builds up. Employees may revert to old methods, create manual work-arounds, or simply disengage from the transformation effort.

This cultural friction can undermine even the most technically sound implementation, leading to low productivity, high turnover, and a legacy of cynicism that makes future change initiatives even harder. In short: Software-first thinking digitizes dysfunction and demoralizes your workforce.

So, how does an organization avoid the pervasive traps of software-first thinking and ensure that strategy explicitly drives every aspect of its SAP implementation? It requires disciplined action from leadership before you ever touch a configuration screen. It starts by defining and publicizing the strategic North Star. Be absolutely crystal clear on your top two to three overarching business priorities that this SAP implementation will directly enable. Examples include "become a real-time enterprise," "achieve global financial consolidation and visibility," "deliver unparalleled customer service through integrated data," or "streamline supply chain to reduce time-to-market by 30 percent." These priorities must be the guiding light for every design decision. Write them down, simplify them, and communicate them relentlessly across the entire organization.

Next, the key is to engage executives deeply and consistently. Do not, under any circumstances, delegate the strategic design and decision-making for SAP to the IT department alone. Your CFO, COO, head of Sales, chief Supply Chain officer, and other relevant business leaders must be fully engaged and explicitly own the business design decisions. They are the ultimate process owners and must be empowered to make the tough trade-offs that align the system with strategy. Their consistent presence and active participation in key workshops, design reviews, and governance meetings are nonnegotiable.

Documenting strategic principles and using them as tiebreakers is the next step for leadership. Translate your overarching business strategy into a set of concise, actionable design principles. These principles serve as objective criteria for resolving conflicts and making choices during system design. Examples include: "We will always prioritize global process consistency over local customization unless legally required," "Customer experience will consistently outweigh internal convenience in process design," and "We will adopt standard SAP functionality unless a proposed customization delivers clear, quantifiable competitive advantage." These principles become the nonnegotiable "rules of the road" for the project team.

Additionally, test design choices against real business scenarios. As part of the design and blueprinting phase, don't just validate configurations against theoretical requirements. Develop and "walk through" actual complex business scenarios (e.g., "new customer onboarding," "complex order fulfillment," "global financial close") using the proposed SAP design. Does it truly support your growth plan? Does it improve your customer commitments? If not, be prepared to realign the design or challenge the underlying assumptions. This rigorous scenario-based testing uncovers strategic misalignments early.

Next, communicate the "why" relentlessly and empathetically. Every employee, from the executive floor to the operational front lines, should understand how the SAP project directly connects to the overall business strategy and what tangible benefits it will bring to them and the company. Without this continuous, empathetic communication, SAP feels like "extra work" or an imposed burden rather than the enabling force for shared organizational goals. Highlight specific improvements relevant to different functions.

Lastly, prioritize process reengineering over customization. Before considering any customization, exhaust all possibilities for adapting your existing business processes to fit standard SAP functionalities. View SAP's embedded best practices as an opportunity to simplify, optimize, and standardize your operations rather than a system that must be bent to accommodate every existing nuance. Customization should be the absolute last resort, justified by an undeniable strategic imperative.

THE LEADERSHIP IMPERATIVE: YOUR RESPONSIBILITY, YOUR POWER

At the absolute heart of a successful strategy-over-software approach lies courageous and disciplined leadership. Only the executive team can define the overarching business vision. Only executives possess the authority and strategic perspective to align the organization's efforts around that vision. And only executives can make the tough, often uncomfortable, trade-offs and decisions that ultimately shape the new SAP system and determine its effectiveness as a strategic asset.

SAP consultants and system integrators are invaluable partners. They bring deep product expertise, implementation methodology, technical know-how, and additional human capital. They can advise, guide, configure, and develop. But they cannot, and should not, dictate who your company wants to be, what its strategic goals are, or what its future operating model should look like. That profound responsibility for strategic direction is a leadership imperative that absolutely cannot be outsourced or delegated.

When leaders abdicate this core role, allowing the project to become a mere technical exercise, SAP becomes a commodity, a cost center, and a source of internal frustration. When leaders courageously embrace this role, providing clear vision and strategic direction, SAP transforms into a powerful strategic weapon—a catalyst for sustained growth, innovation, and competitive advantage.

Here lies the profound paradox of successful SAP implementations: When these projects succeed spectacularly, it is rarely because of SAP alone. It is because visionary leadership meticulously leveraged SAP as a powerful amplifier for their already clear and compelling business strategy. Conversely, when these projects fail, it is almost never truly because of SAP's inherent limitations. It is, instead, because there was no clear, owned, and consistently communicated strategy for SAP to amplify in the first place. The software, lacking strategic direction, simply reflected and exacerbated existing organizational weaknesses.

SAP is not the star of your business transformation. Your meticulously defined, passionately owned, and relentlessly executed strategy is the undisputed star.

Think of SAP as the magnificent stage, the sophisticated lighting, the crystal clear sound system, and the cutting-edge special effects. These are all indispensable elements for a world-class performance. But the play itself—the compelling story your business tells the world, the drama of its operations, the narrative of its innovation, the arc of its customer relationships—that comes entirely from you, the leaders. Without a powerful, well-written script and a clear vision for the performance, even the most technologically advanced stage in the world will leave your audience (your employees, your customers, your stakeholders) profoundly disappointed.

Get the strategy right, and SAP will make it shine brilliantly.

5

Master Data or Disaster Data?

When executives, project managers, and even seasoned IT profession-als envision an SAP implementation, their minds almost invariably gravitate toward the dynamic aspects of business processes: the seamless flow of order-to-cash, the efficiency of procure-to-pay, or the precision of plan-to-produce. They focus on intricate workflows, high-volume transac-tions, intuitive dashboards, and the promise of extensive automation. Yet amid this flurry of operational and technological aspirations, few pause to genuinely consider, or adequately prioritize, the single, foundational ele-ment that quietly underpins and enables every one of these functions: data.

Specifically, we are talking about master data. Master data represents the core, non-transactional entities that define your business: your cus-tomers, your vendors, your materials (products and services), your pricing structures, your bills of materials, your chart of accounts, and critical orga-nizational units. These are not merely records; they are the fundamental building blocks, the atomic particles, of every single transaction and busi-ness process that occurs within your enterprise. Without master data that is impeccably clean, consistently structured, and rigorously governed, your shiny new SAP system—regardless of its power, configurability, or "intelligence"—will be rendered little more than a very expensive, technologically advanced machine that merely produces garbage at scale.

There is an immutable, simple, yet brutally unforgiving rule that every company embarking on an SAP journey must learn, internalize, and adhere to without exception: If your data is broken, your business will be broken in SAP.

WHY MASTER DATA MATTERS MORE THAN YOU THINK

SAP, by its very design, is an unforgiving system when it comes to master data. Unlike the flexibility (or, perhaps, tolerance for sloppiness) offered by disparate spreadsheets, disconnected legacy systems, or manual workarounds, SAP forces a profound level of discipline and standardization. Its integrated architecture means that a single error or inconsistency in a master data record will not remain isolated; it will ripple through the entire system with immediate, and often painful, operational and financial consequences across multiple modules and processes.

Let's illustrate this with concrete examples of how master data imperfections translate into real-world business disruptions. If your customer master data is plagued by duplicates (e.g., "IBM Corp," "Intl. Business Machines," "IBM Co."), outdated addresses, or inconsistent payment terms, the downstream impacts are severe. Invoices may fail to be generated or sent to the correct entity, leading to revenue leakage. Shipments may be delayed or delivered to the wrong location, causing customer dissatisfaction and re-shipping costs. Collections efforts become messy and inefficient, straining cash flow and requiring manual reconciliation. Sales and marketing efforts become untargeted, sending duplicate promotions or irrelevant offers, eroding brand perception.

The material master is the heart of manufacturing, inventory, and sales. If your material master data is incomplete (e.g., missing weights, dimensions, or hazardous material flags), inconsistent (e.g., different units of measure for the same item across plants), or inaccurate (e.g., incorrect safety stock levels), the consequences are immediate. Production orders may not run correctly due to missing component data. Product costing will be wildly inaccurate, leading to incorrect pricing and distorted profitability analysis. Inventory balances will be unreliable, causing stockouts, excess inventory, and inefficient warehousing. Supply chain planning models will generate flawed forecasts.

Accurate vendor master data is crucial for procurement and finance. If your vendor addresses are outdated, banking details are incorrect, or supplier IDs are duplicated, the risks are substantial. Payments may go to the wrong account, causing payment delays, strained supplier relationships, and even potential fraud. Critical supplies may be delayed if purchase orders cannot be processed correctly. Supplier performance metrics become impossible to track accurately, hindering strategic sourcing.

Financial master data discrepancies, which include elements like the chart of accounts, cost centers, profit centers, and general ledger accounts, can lead to inconsistent financial reporting, misaligned performance metrics, and significant challenges during system integration or consolidation. If these are riddled with redundant codes, inconsistencies in hierarchy, or incorrect assignments, the impact on financial integrity is severe. Financial reporting will be distorted, leading to flawed business decisions based on inaccurate profitability or departmental performance. Audits become nightmares, as reconciliation requires extensive manual effort. Regulatory compliance can be jeopardized if transactions are misclassified.

The bottom line is undeniable: Master data is not "back-office administration" or a trivial IT task. It is the lifeblood, the central nervous system, and the literal bloodstream of your SAP system. If this bloodstream is contaminated, the entire body of your business operations will suffer, exhibiting symptoms ranging from minor operational irritations to catastrophic financial and reputational damage.

THE DATA ILLUSION: "WE'LL CLEAN IT LATER"

One of the most pervasive, yet ultimately fatal, mistakes companies make when embarking on an SAP implementation is the misguided assumption that they can "fix the data later." During the initial phases of an SAP project, the allure of diving into system design, blueprinting new processes, and discussing automation possibilities is overwhelmingly strong. Data cleansing, by contrast, feels incredibly tedious, unglamorous, and, in the eyes of many, secondary to the exciting work of building the new system. It's often viewed as a chore, not a critical path activity.

But here's the brutal, often project-killing truth: Data migration is precisely where a staggering number of SAP projects falter, get delayed indefinitely, or completely die. It is the silent assassin of project timelines and budgets.

I've personally witnessed projects suffer crippling delays of many months—sometimes over a year—because foundational elements like customer records could not be reconciled across fragmented legacy systems, or because product hierarchies were fundamentally inconsistent across different regional business units. I've seen go-lives, meticulously planned for months, nearly collapse entirely during the final cutover weekend because payment terms were not standardized, or because the new chart of accounts was riddled with redundant, unmapped legacy codes, preventing critical financial postings.

Too often, leaders and project teams alike tragically underestimate the true scale, complexity, and sheer effort required for comprehensive data cleansing and migration. They might assume that data cleansing will take a mere few weeks of focused effort when, in reality, for a complex enterprise, it can easily consume many months, or even a year or more, of dedicated resources and cross-functional collaboration. And if the initial project timeline and budget do not account for this monumental effort, integrating it from the very beginning, then disaster inevitably follows. This oversight creates a bottleneck that stifles all downstream activities, leading to escalating costs, missed deadlines, and severe blows to project morale and stakeholder confidence.

A CASE IN POINT: THE UNFORESEEN DATA QUAGMIRE

Let me share a vivid example that powerfully illustrates this point. A client of ours, a multibillion-dollar global manufacturing company with operations across three continents, was embarking on a critical SAP S/4HANA implementation. During the initial planning phase, their internal teams, with genuine but misplaced confidence, assured us that their master data was "in pretty good shape." They acknowledged some inconsistencies but believed they were minor.

As we began the meticulous preparation for SAP data migration, however, the stark reality emerged in the following areas of their business:

- Customer records: Our data profiling tools revealed that over 40 percent of their customer records were duplicates, spread across different legacy sales systems, often with slight variations in naming conventions or addresses. This meant a single physical customer might appear multiple times, distorting sales reports, complicating invoicing, and making a unified customer view impossible.

- Vendor data: Their vendor addresses and banking details were significantly outdated. Many suppliers were listed under multiple slightly different IDs due to decentralized procurement processes, leading to redundant contracts and fractured supplier relationships. In some cases, payments were still being sent to defunct bank accounts.

- Material masters: The material masters (their product catalog) were a chaotic mess. Different manufacturing plants and sales regions used entirely different units of measure for the exact same item, leading to confusion in inventory, production planning, and sales order entry. Product descriptions varied wildly, and bills of materials were incomplete or inconsistent across product lines.

The effort required for comprehensive data cleansing, de-duplication, standardization, and enrichment became a project of staggering proportions. It involved dedicated business resources from Finance, Sales, Procurement, and Operations, working closely with data specialists for nearly nine months—a period that ultimately took longer than the core design and build phases of the actual SAP system configuration. What the client initially perceived as a relatively minor "IT data issue" blossomed into a major cross-functional business transformation challenge that demanded significant executive attention and resource allocation.

The lesson from this, and countless other projects, was chillingly clear: Master data readiness is not a secondary task; it is a primary determinant that can single-handedly make or break your entire SAP timeline and budget. Ignoring it is akin to building a skyscraper on a swamp.

Another dangerous misconception is the persistent belief that master data management is solely the purview of the Information Technology (IT) department. This notion is fundamentally wrong—and incredibly dangerous to your SAP implementation and ongoing business health. While IT plays an absolutely critical role in providing the tools, infrastructure, and technical support for data management, the accountability and ultimate ownership of master data unequivocally reside with the business functions.

Why is this a business responsibility? Because master data reflects the very essence of how your business operates, sells, produces, and finances itself. Only the business truly understands the nuances of a customer hierarchy for sales reporting, the critical attributes required for a material in a manufacturing bill of materials, the correct payment terms for a specific vendor relationship, and the accurate classification of a financial transaction for regulatory reporting.

If the business does not actively take ownership of defining, creating, and maintaining master data, IT will be forced to make educated guesses or default to technical considerations. And those guesses will rarely, if ever, align perfectly with operational reality or strategic business needs. The inevitable result is ongoing friction, constant rework, a pervasive lack of trust in the system's output, and a system that fails to meet user expectations.

The most successful SAP projects I've witnessed establish robust data governance councils or boards early in the project life cycle. These cross-functional bodies, typically led by senior business leaders with strong IT representation, ensure that data is treated as a strategic asset, not an afterthought. For example, a CFO may oversee financial data, while the head of Supply Chain or Sales takes responsibility for material, vendor, or customer data.

The council's role extends beyond oversight. They define clear, enterprise-wide standards for how data is named, formatted, and validated, ensuring consistency across every corner of the business. They assign ownership by designating senior leaders as Data Owners, accountable for the strategic quality and integrity of their domains, and appointing Data Stewards, who are responsible for the day-to-day accuracy and integrity of information. When conflicts inevitably arise—whether over definitions, approving data quality initiatives, or prioritizing data remediation efforts—the council has the authority to make critical decisions that keep the project moving forward.

Equally important, they establish data life cycle processes for how data is created, used, updated, archived, and eventually retired, ensuring its integrity throughout its lifespan. To close the loop, they implement metrics and dashboards to continuously monitor data quality, identify anomalies, and track improvement efforts.

By treating master data not as a static, one-time project task but as a strategic asset requiring ongoing governance, organizations elevate its importance and embed the discipline necessary for long-term SAP success.

THE HIDDEN COSTS OF BAD DATA: A SILENT DRAIN ON PROFITABILITY

Let's be unequivocally clear: Bad data is outrageously expensive. Its costs are often hidden and insidious and accumulate over time, manifesting as revenue leakage, crippling operational inefficiency, heightened compliance risk, and pervasive customer dissatisfaction. These aren't just minor inconveniences; they are significant drains on your profitability and competitive edge.

Consider just a few of the tangible, yet often unseen, hidden costs:

- Revenue leakage: A duplicate customer record might lead to sales teams inadvertently extending duplicate discounts or offering promotions to the same customer multiple times, directly eroding profit margins. Incorrect pricing master data can result in underbilling.

- Operational inefficiency: An incorrect material weight or dimension may cause over-shipment costs, penalties with carriers due to miscalculated loads, or inefficient warehouse utilization. Inaccurate inventory counts lead to unnecessary emergency orders, expedited shipping fees, or lost sales due to perceived stockouts. Poor data leads to manual work-arounds, reconciliation efforts, and firefighting, consuming valuable employee time.

- Compliance risk and audit failures: A misclassified general ledger (GL) account or inconsistent financial master data can severely distort financial reporting, leading to noncompliance with accounting standards, regulatory bodies, or tax laws. This can trigger significant fines, reputational damage, and complex, time-consuming audits that require immense manual effort to correct.

- Strained relationships and reputational damage: An outdated vendor record might trigger late payments, damage long-standing supplier relationships, or even expose the company to fraud if payments are diverted. Inaccurate customer data leads to incorrect orders, delayed deliveries, or poor customer service interactions, eroding trust and loyalty.

- Flawed strategic decisions: Poor data quality at the source translates directly into unreliable reports and dashboards in SAP. Leaders relying on this flawed information will make suboptimal business decisions regarding market entry, product development, pricing strategies, or resource allocation, ultimately hindering growth and competitiveness.

- Wasted marketing efforts: Inaccurate or duplicate customer data leads to marketing campaigns being sent to the wrong people, to the same person multiple times, or to nonexistent contacts, wasting precious budget and potentially irritating customers.

Companies often only discover these mounting costs much too late—typically after go-live, when customers begin to complain vociferously, critical audits fail spectacularly, supply chains falter repeatedly, or internal reporting inconsistencies become undeniable.

By then, the cost of remediation is astronomical.

Here lies the profound paradox: While neglecting master data is a direct path to disaster, the inverse is equally true. When master data is meticulously clean, consistently maintained, and strategically governed, it ceases to be merely an operational hurdle and transforms into an incredibly powerful transformation lever and a sustainable competitive advantage. Good data unlocks new levels of insight, automation, and intelligent decision-making that propel the business forward.

Advanced analytics and insights provide reliable customer and product hierarchies, which, combined with accurate financial and operational data, make reporting genuinely meaningful. Suddenly, you can confidently analyze profitability by customer segment, geographical region, product line, or distribution channel with unprecedented confidence. This empowers data-driven strategic planning and allows for targeted interventions to improve performance. Clean data is the prerequisite for any successful AI/ML initiative or predictive analytics.

Seamless automation creates standardized vendor records enable fully automated invoice processing and payment workflows. Clean material masters allow for automated replenishment planning and intelligent manufacturing scheduling. Automated processes eliminate manual errors, reduce operational costs, and free up human capital for higher-value activities.

Accurate, real-time customer data (including order history, service interactions, and preferences) enables consistent pricing, faster and more accurate order processing, and highly personalized customer service. This directly translates into increased customer satisfaction, loyalty, and repeat business.

Clean and consistent financial master data ensures that tax reporting, audit trails, and regulatory compliance are airtight. This reduces exposure to fines, legal challenges, and reputational damage. It provides financial transparency and control that are critical for public companies and regulated industries.

A clean, standardized master data foundation significantly simplifies mergers, acquisitions, and divestitures. Integrating new entities onto a common data model is faster and less risky. It also enables quicker expansion into new markets or the rapid launch of new products because the underlying data structures are robust and adaptable.

Having a single source of truth means good master data that enables SAP promises. All departments operate from the same consistent set of core business entities, eliminating data silos, reconciling disparate reports, and fostering cross-functional collaboration.

In essence, master data is not just an implementation hurdle to be cleared; it is a fundamental enabler of business intelligence and operational excellence and, ultimately, a significant source of competitive differentiation when handled with the strategic importance it deserves.

PRACTICAL STEPS TO GET DATA RIGHT

So, how can you effectively navigate this critical path and avoid turning "master data" into a catastrophic "disaster data" scenario? It requires a disciplined, proactive, and cross-functional approach, with strong executive sponsorship:

1. Start early—the moment you commit to SAP: Begin data cleansing and preparation efforts the very instant your organization makes the strategic decision to move forward with SAP. Do not, under any circumstances, wait until the system-build phase. This often means running a parallel "data program" that precedes and then runs concurrently with the main SAP implementation project.

2. Assess honestly—perform a data health check: Engage specialists (internal or external) to perform a rigorous, objective data health check or data profiling of your existing master data across all legacy systems. Quantify duplication rates, completeness, consistency, and adherence to any existing (or desired) standards. Don't assume your data is "good enough"; prove it with data.

3. Establish business ownership—not just IT accountability: Formally assign Business Owners to each critical master data domain (e.g., CFO for financial master data, VP Sales for customer data, VP Operations for material data, CPO for vendor data). These leaders are ultimately accountable for the quality and strategy of their data. Below them, identify and empower Data Stewards— operational employees who are responsible for data quality in their daily work.

4. Define enterprise-wide standards—data harmonization and standardization: Work cross-functionally to define clear, enterprise-wide standards for all critical master data elements. This includes

   a. Naming conventions (e.g., consistent formats for customer names, product descriptions)

   b. Units of measure (e.g., standardizing "EA" vs. "PC" vs. "Each")

   c. Payment terms, shipping conditions, and other critical business attributes

   d. Hierarchies (e.g., customer groups, product categories, GL account structures). This often involves data harmonization across disparate regional or historical systems.

5. Invest in data tools and expertise: Do not rely solely on manual Excel-based fixes for large-scale data cleansing. Invest in modern data profiling, data quality, data migration, and data governance tools. These solutions can automate much of the heavy lifting, identify patterns, enforce rules, and accelerate the process significantly. Augment your team with specialized data architects and data quality analysts.

6. Plan for ongoing governance—beyond go-live: Understand that master data discipline is not a one-time project. Create a sustainable data governance model that extends far beyond the go-live date. This includes establishing a standing data governance council, defining roles for ongoing data maintenance, and implementing continuous data quality monitoring processes. SAP requires ongoing discipline to keep data clean.

7. Prioritize critical data—phased approach to perfection: Not all data is equally critical for go-live. Prioritize the master data that directly drives core transactional processes (e.g., active customers, frequently purchased materials, key vendors, essential financial accounts). Focus on getting this foundational data perfect for your

initial go-live and then plan for subsequent waves of data cleansing for less critical or historical data. An incremental approach can be more manageable.

8. Automate data creation and maintenance where possible: Look for opportunities to automate the creation and maintenance of master data through integration with other systems (e.g., CRM for customer creation, PLM for material creation) or through standardized templates and workflows within SAP.

THE HUMAN SIDE OF DATA: CULTURE, COLLABORATION, AND ACCOUNTABILITY

It's easy, and often tempting, to conceptualize master data as a purely technical challenge—a task for databases and IT specialists. But in reality, master data quality is deeply, inextricably human and cultural. Every duplicated customer record often reflects a lack of cross-functional collaboration between sales teams or inconsistent entry processes. Every inconsistent material code points to organizational silos between different plants or business units. Every redundant vendor record frequently exposes breakdowns in procurement processes or a lack of centralized authority.

Therefore, genuinely fixing your master data means far more than just running cleansing scripts; it means profoundly fixing the way people work together. It demands alignment across business functions so that everyone agrees on data definitions, standards, and processes. It requires clear ownership with responsibilities and accountabilities for the accuracy and integrity of each data element throughout its life cycle. Collaboration is essential, fostering a culture where departments actively work together to maintain data quality, understanding its ripple effects across the enterprise.

It also calls for accountability, through performance metrics, audits, or other mechanisms to hold individuals and teams accountable for data quality. Finally, it requires change management for data, recognizing that changing how people interact with data is a significant behavioral shift. This requires targeted communication, training, and support to help users understand the "why" behind data quality and how it impacts their daily work. Incentivize good data practices.

That's why mastering master data is not just a systems exercise—it is a fundamental organizational maturity exercise that reveals and addresses deeper issues of process discipline, cross-functional collaboration, and shared responsibility.

When SAP projects unfortunately fail to deliver their promised value, the blame is often misdirected. People tend to point fingers at the software itself, the consulting firm, the unrealistic timeline, or the project management. Rarely, if ever, do they first point to the actual silent culprit: dirty, inconsistent, poorly governed, and unowned data. Yet, in my extensive experience, master data is consistently the single greatest source of hidden risk in every SAP project, far more so than technical configuration challenges or integration complexities.

The good news, however, is that this hidden risk also represents the single greatest lever for success and competitive advantage. Companies that possess the foresight and discipline to invest early in data assessment and cleansing, that establish robust data governance, and that empower business leaders to truly own their data as a strategic asset not only mitigate catastrophic risks—they achieve profound benefits. They build a resilient foundation for long-term operational efficiency, superior analytics, seamless automation, and sustainable competitive differentiation.

So, as you embark on your SAP journey, pause and ask yourself this critical question with unflinching honesty: Is your data truly ready for SAP? Or will your ambitious project, despite its immense investment, be remembered not as a story of business transformation but as a cautionary tale of disaster data? The choice, and the responsibility, is entirely yours.

6

Best Practices Are a Myth: Finding the Right Fit in SAP

When executive leadership teams embark on the journey of an SAP implementation, they are often met with a highly seductive and seemingly reassuring sales pitch: "With SAP, your company will be adopting industry 'best practices.' We'll take what works for the world's leading companies, the gold standards of operational excellence across the globe, and apply those proven methodologies directly to your organization."

On the surface, this promise sounds incredibly appealing. Who wouldn't want to benefit from tried-and-true, battle-tested practices that have been refined and validated across thousands of global leaders in various industries? The allure of such a proposition is powerful: Why expend valuable time, resources, and intellectual capital trying to reinvent the wheel when SAP, through its extensive research and development, has ostensibly already captured the "right way" to run critical functions like finance, sales, manufacturing, and supply chain? It offers a comforting sense of de-risking the project, suggesting a ready-made blueprint for success.

But here, a crucial truth must be revealed, a truth often obscured by the persuasive rhetoric of templates and accelerators: There is, in fact, no such thing as a truly universal, one-size-fits-all "best practice" that applies equally and optimally to every company, in every industry, under all market conditions.

What might represent a "best practice" for a high-volume, commoditized consumer goods manufacturer operating in a stable market could very well be a catastrophic, value-destroying process for a nimble, engineer-to-order aerospace company navigating bespoke client demands. "Best practices" are not akin to the immutable laws of physics or universally applicable mathematical theorems. They are, at their core, simply patterns, configurations, or methodologies that have proven effective somewhere, for someone, at a specific point in time, and within a particular context. To blindly assume their universal applicability is a dangerous oversimplification.

SAP's extensive reference processes, preconfigured content, and industry solution maps are undeniably built upon patterns and aggregated commonalities observed across its vast global customer base. These represent efficient, often legally compliant, ways of structuring operations. However, it's imperative to understand that these are primarily "common practices" or "reference models"—a starting point, a baseline, or a robust menu of options, but definitively not a mandated, singular "best practice" destination.

To illustrate this critical distinction, let's consider a few simple, yet illustrative examples from diverse industries. A global pharmaceutical company operates under stringent regulatory requirements (e.g., FDA, EMA). Their inventory management processes demand meticulous, validated procedures. Every single batch of raw material, works-in-progress, and finished goods must be fully traceable from supplier to customer. Every transaction must be documented with absolute precision for audit and compliance. Their "best practice" for inventory might emphasize rigorous batch management, detailed quality inspection processes, and extensive audit trails within SAP.

The second company, a small, fast-moving consumer electronics distributor, on the other hand, operates on razor-thin margins and in a highly competitive market where speed and flexibility are paramount. They prioritize rapid stock turnover, minimal warehousing costs, and efficient pick-pack-ship operations. Their "best practice" for inventory is likely centered on streamlined goods receipt, simplified stock movements, and real-time availability checks, with less emphasis on granular batch traceability (unless legally required for specific product types).

If both of these companies were to blindly adopt the exact same SAP "best practice" inventory process—say, the highly detailed, regulated one— the electronics distributor would drown in crippling inefficiency, unnecessary overhead, and a loss of their competitive agility. Conversely, if the pharmaceutical company adopted the distributor's agile-focused process, they would almost certainly fail critical regulatory audits, risking massive fines and reputational damage. The "best practice" for one is a "worst fit" for the other.

Let's look at how pricing strategy "best practices" can be different for various companies as well. For example, a utility company might have a "best practice" pricing process driven by regulatory mandates, long-term contracts, and stable, fixed tariffs. Their SAP pricing configuration would reflect complex rate structures, usage-based billing, and robust compliance checks.

Conversely, a fashion retailer that competes on dynamic pricing and frequent promotions might have a "best practice" focusing on rapid price changes, flash sales, personalized discounts, and complex promotional hierarchies.

Applying one to the other would clearly be disastrous.

What SAP offers as a "best practice" is valuable insight into how many companies operate and often represents an efficient way to configure the system. But it is not a strategic imperative that automatically aligns with your specific business model, customer value proposition, or competitive landscape.

THE PERILS OF BLIND ADOPTION: WHEN "STANDARD" BECOMES A STRAITJACKET

One of the most common and costly mistakes in SAP implementations is the uncritical acceptance and blind adoption of these so-called best practices as gospel rather than treating them as flexible guidance. When companies implement SAP's template processes without rigorously asking the fundamental question, "Does this truly serve our unique business strategy, strengthen our competitive differentiators, and genuinely improve our customer experience?" they run the severe risk of major strategic and operational misalignment.

The consequences of this blind adoption are far-reaching and detrimental. A consumer goods company, for instance, once implemented an SAP "best practice" pricing procedure that was designed for high-volume, undifferentiated retail. Their true market, however, lay in customized pricing models and bespoke contracts for niche, high-value customers. The rigid, template-driven pricing stripped away this competitive advantage, forcing them into a commoditized pricing model that didn't fit their unique value proposition. They rapidly lost market share to more agile competitors who could offer tailored solutions.

Another client, an industrial equipment manufacturer specializing in engineer-to-order (ETO) products, insisted on using SAP's "standard" production planning approach. This template was ideally suited for repetitive, make-to-stock manufacturing environments. It looked efficient on paper but utterly failed to account for the unique complexities of their ETO process, which involved extensive presales engineering, customer-specific bill of materials, and dynamic production schedules. The result was endless manual work-arounds outside the system, pervasive user frustration, crippling delays in production, and, ultimately, costly, complex customizations forced upon the "standard" system to make it barely functional.

Paradoxically, blind adoption of an ill-fitting "best practice" often leads to more customization, not less. When a template doesn't genuinely align with core business needs, the company finds itself retrofitting its unique requirements onto a rigid, predefined structure. This leads to inefficient "Z-code" (custom programs in SAP ABAP), complex enhancements, and convoluted interfaces that try to force-fit a square peg into a round hole. This creates a "Frankenstein system," which is far more expensive to build, test, maintain, and upgrade than a system designed for "best fit" from the outset.

User adoption also suffers. When employees are forced to adopt processes that feel unnatural, illogical, or detrimental to their daily work— simply because they are "best practice"—they quickly become disengaged and resistant. They see the new system as an impediment rather than an enabler, leading to work-arounds, cynicism, and a failure to fully leverage the new capabilities.

Finally, strategic focus gets diluted. When design decisions are dictated by generic templates rather than strategic objectives, the project can lose its way. The focus shifts from solving critical business problems or enabling strategic goals to simply "going live" with a preconfigured system, regardless of its alignment with the company's true North Star.

Ultimately, "best practices" become dangerous when they are treated as a substitute for rigorous strategic thinking, critical analysis, and deliberate design.

Beyond the commercial drivers, there's also a powerful psychological reason why "best practices" are so seductive to executive leadership: They offer a convenient way to relieve executives of ultimate responsibility.

If a complex, multimillion-dollar SAP project later falters or fails to deliver its promised value, it can be psychologically comforting for leaders to say, "Well, we just followed SAP's best practices. We did what the software dictated." This narrative subtly shifts accountability away from the internal leadership team and onto the software vendor or the generic "industry standard." It provides a shield against potential blame.

But here's the stark reality: You cannot simply outsource responsibility for your fundamental business model, your strategic direction, or your company's long-term success. Whether you choose to adopt a standard template, implement a complex custom process, or opt for a hybrid approach, the ultimate decision to do so rests squarely with your leadership team. Blaming "best practices" or "the software" won't help when customers are unhappy, operational margins shrink, or your competitive position erodes. The responsibility for strategic alignment and outcome rests with you.

THE ILLUSION OF SPEED: SHORTCUTS THAT LEAD TO COSTLY DETOURS

Why do so many companies, despite their best intentions, fall headfirst into this "best practice" trap? One of the most powerful underlying reasons is the seductive promise of speed.

Consulting firms, eager to win engagements and demonstrate quick wins, often market "best practices" as a direct path to "accelerated implementation." The sales pitch sounds compelling: "Don't waste months designing processes from scratch, iterating through requirements, and dealing with internal debates. Just adopt what already works! We can get you live with SAP's standard template much faster, saving you time and money."

But here lies a critical paradox: Cutting corners early in a complex SAP implementation usually ends up costing significantly more later, both in terms of time and money. What appears to be a convenient shortcut often creates a far longer, more painful journey marked by extensive rework, pervasive user resistance, spiraling customization costs, and unforeseen operational complexities.

I've observed numerous projects that proudly declared themselves "live with SAP best practices in a record six months!"—only to spend the subsequent two or three years engaged in a painful, costly cycle of customizing, patching, debugging, and desperately retraining users because the initially chosen "best practices" simply did not align with the granular realities of their business operations, their unique market demands, or their deeply ingrained organizational culture. The initial "speed" was an illusion, masking a foundational misalignment that eventually surfaced as massive delays and budget overruns.

True speed in an SAP transformation does not come from skipping the critical design and alignment phases; it comes from designing wisely, strategically, and with a "best fit" mindset up front. This involves investing the necessary time and intellectual capital in understanding your unique business needs, making deliberate design choices, and ensuring that the adopted processes truly serve your strategic goals. A slightly slower start, built on a solid "best fit" foundation, invariably leads to a significantly faster, smoother, and more successful implementation in the long run.

The alternative to blindly adopting "best practices" is not a descent into chaos, nor does it imply reinventing every single process from scratch. That would be equally inefficient and counterproductive. SAP's best practice content and preconfigured solutions certainly hold immense value—but only if they are approached as a comprehensive menu of intelligent options rather than an unchangeable mandate.

The fundamental question that every executive and project leader must relentlessly ask throughout the design phase is not "What does SAP say is the 'best practice'?" but rather "What is the 'best fit' for our business, given our unique strategic vision, our specific customer value proposition, our competitive landscape, and our organizational capabilities?"

This is precisely where leadership must step in decisively. Executives must resist the seductive temptation to abdicate critical design decisions to generic software templates or to the default recommendations of consultants. Instead, they should utilize SAP's best practices as a rigorous benchmark and a rich source of proven patterns. From this benchmark, they must then deliberately and strategically choose what to adopt, adapt, or reject.

Let's break down this crucial framework:

- Adopt: Embrace SAP's standard "best practice" where the process is truly a commodity or undifferentiating for your business. These are processes where there is little to no competitive advantage in reinventing the wheel and where standardization brings efficiency, compliance, or risk reduction.

  - Examples: Standard general ledger postings, basic tax calculation logic, generic procure-to-pay steps for office supplies or indirect materials, standard HR processes like time entry and payroll, basic accounts payable/receivable. In these areas, adopting standard practices saves time, reduces risk, and often ensures compliance.

- Adapt: Modify or tailor an SAP standard "best practice" where the process touches your core differentiators or unique business needs but can still align with SAP's framework with minor, well-justified changes or configuration. This requires careful analysis to ensure the adaptation doesn't introduce excessive complexity or negate future upgradability.

  - Examples: Minor adjustments to a standard sales order flow to accommodate a specific customer approval process, slight variations in a manufacturing routing to handle a unique production step, or tailoring a standard reporting structure to meet specific executive dashboard requirements. These adaptations should be meticulously documented and rigorously tested.

- Reject: Consciously decide to not adopt an SAP "best practice" where it would fundamentally undermine your unique value proposition, introduce unacceptable operational inefficiency, or create significant competitive disadvantage. This is where you might implement a truly custom process (with extreme caution and a strong business case) or choose an alternative SAP solution (e.g., a specific industry solution).

  - Examples: If your entire business model relies on a highly complex, unique pricing algorithm that cannot be reasonably configured within standard SAP, you might reject the standard pricing process and build a custom interface. Or if your product configuration is so bespoke that it requires a specialized configurator, you might reject SAP's standard variant configuration in favor of an integrated third-party tool. Rejection should always be the exception, supported by a clear, undeniable strategic imperative and a thorough cost-benefit analysis.

This deliberate, strategic approach—of consciously choosing to adopt, adapt, or reject—is inherently slower at the initial planning and design stages. It demands more thought, more internal alignment, and more rigorous analysis. However, it invariably saves enormous pain, cost, and frustration later in the project and, crucially, after go-live, ensuring that the SAP system truly becomes an accelerator for your unique strategy, not a straitjacket.

THE CONSULTANT'S ROLE IN THE MYTH: COMPLICITY AND RESPONSIBILITY

Let's be candid: A significant part of the "best practice" myth is, unfortunately, perpetuated by some consulting firms and individual consultants.

For a consultant, it is often far easier, faster, and less cognitively demanding to simply recommend, "Just take the template," or "This is SAP standard, so we should do it this way." Relying on templates and preconfigured solutions can streamline the consulting firm's internal processes, reduce their own project complexity, accelerate their billing cycles, and potentially lower their internal training costs. However, what reduces project complexity for the consultant does not automatically translate into a beneficial outcome or reduced complexity for you, the client.

A truly responsible, ethical, and value-driven consulting firm will actively challenge the "best practice" myth. They will understand that their primary role is to serve your unique business needs and strategic objectives, not to simply implement a generic template. A good consultant will

- Act as a strategic partner: They will engage deeply with your executive team to understand your vision, strategy, and competitive landscape.

- Ask probing questions: They will push back if the vision is vague or inconsistent. They will ask tough, insightful questions: "Yes, SAP offers this as a reference, but does it truly fit your vision? Does it strengthen or weaken your competitive advantage? Have you considered the long-term implications for your specific market?"

- Facilitate the adopt/adapt/reject discussion: They will guide your business users and leaders through the critical decision-making process, presenting the standard options but forcing the internal team to justify deviations based on strategic imperatives, not just legacy habits.

- Provide context and risks: They will explain the benefits and risks associated with both standard adoption and customization, ensuring you make informed decisions.

- Prioritize client value: Their ultimate goal is your long-term success and strategic enablement, not just a quick technical go-live.

Unfortunately, not every consulting firm operates with this level of responsibility and strategic partnership. Many prioritize speed, maximizing billable hours, or achieving quick, visible wins over cultivating genuine, long-term client success. This is precisely why you, as the client's executive leadership, must own the ultimate decision and be discerning in your choice of consulting partner. You must be prepared to challenge their recommendations and demand strategic alignment.

CASE STUDY: BEST PRACTICE GONE WRONG— THE BESPOKE SERVICE MODEL

When a midsize industrial equipment manufacturer escalated serious concerns about SAP's fit for its "unique" processes, I was brought in as SAP Architect to conduct an independent review. The brief was clear: Examine the current implementation, distinguish configuration and governance issues from genuine product limitations, and recommend a pragmatic remediation plan. Their core business model revolved around not just selling machinery but offering highly specialized, long-term service contracts, field engineering support, and custom preventive maintenance agreements that were unique to each client and piece of equipment. This bespoke service model was their primary differentiator and revenue driver.

When they embarked on their SAP implementation, they decided, against initial internal dissent, to adopt SAP's "standard" service management process. On paper, this template looked efficient: It facilitated quick service calls, generic work orders, and basic warranty tracking. However, it was fundamentally designed for a transactional, high-volume service model, not their highly personalized, contract-driven, field-service-intensive operations.

To force their unique business model into SAP's standard template, the client's team, with their consultants, ended up performing massive amounts of custom development: They created over three hundred custom fields to capture unique contract terms and equipment configurations, wrote dozens of complex "Z-programs" (custom ABAP code) to automate bespoke service logic, and invested millions of dollars annually in maintaining this unwieldy custom code. Their engineers spent more time grappling with the system's rigidity and manual work-arounds than focusing on high-value field service.

The client's leadership openly admitted, in hindsight, "If we had taken more time up front to meticulously design our service processes based on our unique value proposition rather than trying to fit into a generic 'best practice' box, we could have avoided years of operational pain, millions in unnecessary customization costs, and a constant struggle with user adoption. The 'best practice' was fundamentally not 'best' for our business." Their strategic differentiator became their biggest operational burden.

To be fair and balanced, it's important to acknowledge that "best practices" (or, more accurately, standard, efficient processes) are not inherently useless. They can be incredibly valuable and should be embraced enthusiastically when applied to undifferentiating, commodity processes within your organization. In these areas, there is little to no competitive advantage to be gained from reinventing the wheel, and standardization offers significant benefits. For example, there is no reason to create a unique process for buying pens and paper when SAP's standard procure-to-pay flow for indirect materials is both efficient and effective. The same is true for tax reporting formats that are dictated by regulatory compliance; adopting standard SAP functionality for tax calculation and reporting formats (such as VAT or sales tax) saves enormous time and reduces risk.

Basic HR processes also benefit from standardization, including time entry, standard payroll calculations, and basic employee master data. Similarly, general ledger postings follow the universal accounting logic for debits and credits, making standard functionality the most effective approach. Even order creation, in its simplest form—such as direct-from-stock orders—is handled with maximum efficiency through SAP's standard order entry processes.

In these contexts, adopting standard practices can indeed save time, reduce risk, ensure compliance, and free up resources to focus on areas that do provide a competitive edge. The key is discerning when a process is a commodity to be standardized versus a differentiator to be strategically designed.

PRACTICAL GUIDELINES: CULTIVATING A "BEST FIT" MINDSET

So, how can you, as a leader, navigate the complexities and avoid the "best practice" trap, ensuring your SAP implementation delivers maximum strategic value? Here are practical guidelines for cultivating a "best fit" mindset:

- Identify and protect your differentiators: Conduct a rigorous internal exercise to ask, "Which business processes or capabilities truly define our competitive advantage? What makes us unique in the marketplace? Where do we genuinely win customers?" These are the areas that demand deliberate, strategic design and where blind adoption of a "best practice" could be fatal.

- Challenge templates and default recommendations: Never assume SAP's preconfigured content or a consultant's initial template is automatically right for you. Use it as powerful input, a starting point for discussion, and a valuable benchmark—but not as an unchallengeable instruction. Force your team and consultants to articulate why a particular standard process makes sense for your business and your strategy.

- Engage business leaders as designers, not just reviewers: Design decisions related to critical processes should be made by those who own the business strategy and the operational outcome, not solely by IT personnel or external consultants. Empower your CFO, COO, head of Sales, and other key business leaders to actively participate in, and lead, the design workshops. They must own the "to be" process.

- Balance standardization with strategic uniqueness: Develop clear criteria for when to standardize and when to allow for strategic uniqueness. The default should be standardization, but with a robust, data-driven business case required to justify any deviation. A "best fit" approach isn't about avoiding all customization; it's about making smart, justifiable customizations that serve a strategic purpose rather than simply replicating legacy inefficiencies.

- Think long term and future-proofing: Choose processes and system designs that will not only meet today's needs but also scale with your future strategy, growth plans, and evolving market demands. A "best fit" today should ideally be adaptable for the "best fit" of tomorrow.

- Invest in process reengineering first: Before even considering customization for a deviation, exhaust all possibilities for process reengineering. Can your existing process be simplified, optimized, or altered to align with standard SAP functionality while still achieving the desired business outcome? Often, the answer is yes, requiring a shift in internal thinking rather than custom code.

The myth of universal "best practices" in SAP is comforting but profoundly dangerous. It falsely suggests that enduring success can be achieved simply by following someone else's prepackaged recipe. But SAP is fundamentally not a recipe; it is a powerful, flexible platform. Your ultimate success, and the true value you derive from your immense SAP investment, depends entirely on how skillfully and strategically you align that platform with your organization's unique vision, its distinct business strategy, its competitive differentiators, and its core aspirations.

So, the next time someone—whether an internal stakeholder or an external consultant—promises you the panacea of "industry best practices" for your SAP project, pause. Challenge them directly. Ask the critical questions: "Best for whom? Best under what specific conditions? Best in what precise market context? And, most importantly, is it truly 'best' for my unique business, my strategic goals, and my customers?"

The companies that achieve transformative success with SAP are not those who blindly adopt generic "best practices"; they are the astute, disciplined leaders who relentlessly pursue "best fit" practices—those meticulously chosen, designed, and implemented solutions that accurately reflect their own distinct identity, amplify their core strengths, and directly enable their strategic competitive advantage.

Because in the end, your business does not win by being average, by simply mirroring industry norms; it wins by being different, by being strategically superior, and by executing its unique vision flawlessly. SAP, when wielded with strategic clarity, is the powerful enabler of that difference.

The Allure of Customization: A Double-Edged Sword

When an organization first encounters SAP, particularly its core ERP functionalities, a common refrain echoes through the halls from business users and departmental leaders: "Can SAP do it the way we do it today?" This is an entirely fair, even natural, question. After all, employees have been operating the business in a certain familiar manner for years, sometimes even decades. Existing processes, often supported by intricate spreadsheets, siloed databases, and bespoke homegrown systems, have evolved organically over time to reflect the unique history, specific needs, and ingrained habits of that particular company. So, when a powerful, standardized platform like SAP arrives, proposing its own structured way of doing things, the immediate and natural human instinct is to seek comfort in the familiar. The urge is to say: "Can't we just tweak the new system to fit us instead of us having to change and fit it?"

And the answer, from a purely technical standpoint, is almost invariably yes. SAP, by design, is an extraordinarily flexible and extensible system.

With sufficient custom code, strategic enhancements, or carefully integrated partner add-ons, you can indeed make SAP perform nearly any specific function or replicate almost any existing process. This inherent flexibility, this vast capacity for adaptation, is simultaneously SAP's greatest strength and, ironically, its greatest danger. It promises the world, but it can, if wielded without immense discipline, lead you down a path of escalating costs, complexity, and ultimately, project failure.

The fundamental problem with excessive customization is not that it's technically impossible; it is that every single customization comes with a compounding, long-term cost. This cost extends far beyond the initial development effort and infiltrates every aspect of your SAP system's life cycle. Beyond the initial development fees for coders and consultants, customizations demand significant ongoing financial outlays. They require continuous testing with every single SAP support pack, patch, and major version upgrade. They necessitate dedicated support personnel who understand the bespoke code. And when a system upgrade (like moving from SAP ECC to S/4HANA) occurs, these customizations often have to be completely rewritten or reverified, which can be astronomically expensive—sometimes more than the original development cost.

There is also a cost in terms of time. Every instance of customization adds significant delays to your project timeline. Developing custom code is time-consuming. More importantly, rigorous testing of that custom code—ensuring it works as intended, doesn't break other standard functionalities, and integrates seamlessly—takes even longer. Debugging and refining custom solutions can become a perpetual bottleneck, pushing out critical go-live dates and frustrating impatient business stakeholders.

Complexity adds yet another layer of cost. The more you customize the core SAP system, the more unique your solution becomes. This significantly increases system complexity. Each customization creates a deviation from the standard, making it harder to troubleshoot issues, train new users, and find readily available support resources (internal or external). This accumulation of custom code and configurations is often referred to as "technical debt"—a metaphor where the "interest" on the debt is the ongoing cost of maintaining, supporting, and upgrading a nonstandard system. The more technical debt you accrue, the harder and more expensive it becomes to innovate or adopt future SAP capabilities.

Finally, there is a cost of innovation blocking. SAP constantly releases new functionalities, industry solutions, and technological advancements (e.g., AI/ML capabilities, new Fiori apps). If your core system is heavily customized, adopting these new innovations often becomes challenging, if not impossible. The new standard functionality might conflict directly with your custom code, requiring extensive rework of your bespoke solutions or forcing you to forgo the new capabilities entirely. This leaves you operating on an increasingly outdated platform, despite having invested in a modern ERP.

I've witnessed numerous companies whose SAP systems became unique snowflakes due to hundreds, or even thousands, of "Z-programs"—custom-coded enhancements directly modifying standard SAP processes or functionalities. While these worked for them in the short term, addressing specific legacy requirements, they became a crippling liability. When the time came to upgrade their SAP ECC system to SAP S/4HANA, particularly to a cloud-based version, they were effectively stuck. Their "SAP" was no longer a standard, supported product; it was a bespoke Frankenstein system that only a handful of internal experts—or the original, highly expensive consultants—could truly understand and maintain. This significantly increased their total cost of ownership (TCO) and delayed their digital transformation by years.

This, then, is the core of the customization trap: the false, seductive belief that you can implement a powerful, integrated ERP system like SAP and simultaneously keep everything exactly as it was without incurring crippling long-term costs and sacrificing future agility.

WHY COMPANIES CUSTOMIZE TOO MUCH

The question isn't whether customization is possible but why so many organizations fall into this trap despite the well-documented risks. There are several powerful, often interconnected, reasons:

- Resistance to change (the path of least resistance): This is perhaps the most common driver. Business users and departmental managers have ingrained habits and comfort with their existing processes, spreadsheets, and homegrown tools. These established workflows, even if inefficient, feel "safe." Customization feels like the path of least resistance—an easy way to avoid the often uncomfortable work of process reengineering and adapting to new ways of working. Leaders, sometimes lacking the courage or conviction to enforce change, acquiesce to these requests, believing it will smooth over immediate friction.

- Misunderstood or undisciplined requirements gathering: Consultants, and internal project teams, sometimes take business requests at face value without rigorous challenge. Instead of asking, "Why is this truly needed?" or "Can this business outcome be achieved with a standard SAP process or a simpler approach?", they simply document the request as a "requirement" that must be delivered. This lack of critical inquiry and a "fit-to-standard" mindset allows every individual preference to become a perceived "gap" that must be filled by custom development.

- Short-term thinking ("just this one gap"): Leaders and project managers, under pressure to meet immediate operational needs or address specific pain points, often believe that a particular customization will "just solve this one small gap" without fully recognizing the compounding, long-term costs. Each "small gap" adds a piece of custom code, and one becomes ten, ten becomes a hundred, and before long, the project scope balloons exponentially, its complexity spirals, and the original strategic goal of standardization and integration is completely lost in a sea of bespoke development.

- Fear of lost functionality or competitive edge: Companies sometimes customize because they genuinely fear that adopting standard SAP processes will mean losing a unique capability that they believe provides a competitive edge or that it will prevent them from achieving a critical business outcome. This fear, while sometimes legitimate (and we'll discuss justified customization later), often stems from an incomplete understanding of SAP's full capabilities or an unwillingness to adapt internal processes to leverage the system's strengths.

- The "our business is unique" fallacy: A common refrain in customization-prone organizations is "but our business is unique." While every company has unique aspects, very few are so fundamentally different that they cannot leverage 80–90 percent of standard SAP functionality. Often, this "uniqueness" is simply a euphemism for outdated, inefficient, or internally siloed processes that should be standardized for greater efficiency. Leaders must challenge this fallacy.

- Consultant complicity: Unfortunately, some consulting firms may inadvertently (or even deliberately) enable excessive customization because it translates directly into more billable hours and higher project budgets for them, without necessarily prioritizing the client's long-term health.

The interplay of these factors creates a powerful gravitational pull toward customization. It's a comfortable, seemingly quick fix that bypasses the harder work of true transformation and process reengineering.

THE EXPONENTIAL COST OF CUSTOMIZATION: PAYING FOREVER

Let's be brutally clear: Customization is not just expensive up front; it's exponentially expensive forever. It's the gift that keeps on taking, draining resources and inhibiting agility for the lifespan of your SAP system.

The first burden is ongoing maintenance. Every piece of custom code, every bespoke report, every modified interface requires dedicated maintenance. It must be periodically reviewed, updated, and tested with every single SAP support pack, patch, and version upgrade. This is not a trivial task; it can consume significant internal IT resources and external consulting fees annually, diverting funds from value-added innovation.

The most crippling long-term costs are innovation blocking and upgrade pain. If your core code is heavily modified, these new innovations often cannot be adopted easily, or at all, if they conflict directly with your custom developments. The sheer volume of accumulated customizations has, in fact, bankrupted SAP upgrade projects—particularly the move from older ECC systems to S/4HANA—because the cost and complexity of migrating or rewriting decades of bespoke code become prohibitive. SAP itself has publicly warned customers that excessive legacy customization is the number one barrier to adopting SAP S/4HANA Public Cloud and leveraging its full benefits, as the cloud environment has very limited tolerance for core modifications.

Heavily customized SAP systems often lead to a dangerous dependency on one or two key internal or external developers who possess the arcane knowledge of your bespoke programs and unique configurations. This talent dependency can lead to vulnerabilities. If these individuals leave the organization, you are left in an incredibly vulnerable position, struggling to understand, maintain, or evolve your critical ERP system.

Another cost is an increased debugging and issue resolution time. When a problem arises in a heavily customized system, troubleshooting becomes exponentially more complex. Is the issue a standard SAP bug, an error in custom code, or a conflict between multiple customizations? Pinpointing the root cause is much harder and takes longer, leading to extended downtime and operational disruption.

When you deviate significantly from standard SAP, the ability to leverage standard SAP support channels or common knowledge bases among consulting partners diminishes. Your issues become unique to you, requiring specialized, and often more expensive, assistance.

CASE STUDY: THE BANK WITH TWO THOUSAND Z-PROGRAMS—A DEBT SO DEEP

I recall vividly a project with a prominent global bank that had been running SAP ECC for over two decades. Over that incredibly long lifespan, driven by relentless requests from diverse business units and a culture of accommodating every ask, their IT team had consistently said yes to customization. Every new report, every unique workflow, every bespoke process variation was meticulously coded into their system.

By the time the bank began to seriously consider a strategic move to SAP S/4HANA, a crucial next-generation platform, their technical assessment revealed a staggering and utterly crippling discovery: They had accumulated over two thousand custom programs (Z-programs), along with thousands of custom tables, screens, and modifications directly to SAP standard code. These customizations, originally implemented to solve specific problems, had effectively turned their ERP into a highly individualized, proprietary system.

The projected cost of migrating, validating, and adapting these two-thousand-plus custom programs to the new S/4HANA environment was, unbelievably, estimated to be more than the entire original implementation cost of their SAP ECC system twenty years prior. The bank was faced with a stark, agonizing choice: Spend hundreds of millions of dollars migrating obsolete processes and technical debt that provided little strategic value or undertake a complete process redesign and a "greenfield" (new implementation) approach to S/4HANA, essentially paying twice for their core ERP because of the decades of unchecked customization. They ultimately chose the latter, but the financial and organizational pain was immense, a direct consequence of their past customization trap.

The underlying issue driving excessive customization is often a fundamental mindset problem. Too many executives and business users approach an SAP implementation with the wrong backward-looking question: "How do we make SAP work exactly like our old system?" This immediately sets the stage for resistance to change and an insatiable appetite for customization.

But the truly transformative question, the one that unlocks SAP's full potential, is fundamentally different and forward looking: "How can SAP help us work in a better, more efficient, more agile, and more insightful way than ever before?"

If your primary goal is to meticulously replicate the past—to simply digitize existing inefficiencies and deeply ingrained habits—then customization will present itself as the alluringly easy solution at every turn. You will be tempted to build a digital twin of your legacy system, complete with all its flaws. However, if your overarching goal is genuine business transformation, strategic advantage, and future-proofing, then you will be far more inclined to embrace standard processes wherever possible, challenging the status quo and leveraging SAP's embedded best practices as an opportunity for improvement.

WHEN CUSTOMIZATION IS JUSTIFIED: THE EXCEPTION, NOT THE RULE

It's crucial to reiterate that all customization isn't inherently evil or always to be avoided. There are, indeed, legitimate and strategically justifiable reasons to extend SAP's capabilities. When these conditions are met, a carefully designed and managed customization can be a necessary enabler of business value. These exceptions typically fall into a few key categories.

The first is regulatory compliance. When a specific law, industry regulation, or statutory reporting requirement mandates a unique report, a specific data capture, or a control mechanism that is demonstrably not covered by standard SAP functionality, then a customization may be necessary. This must be a clearly documented, nonnegotiable legal requirement, not merely a preference.

Another category is true competitive differentiation. If your unique business model, core intellectual property, or primary competitive advantage genuinely relies on a process or capability that SAP's standard functionality cannot provide, and this unique capability directly drives significant, measurable business value (e.g., market share, profit margin, customer loyalty), then a customization might be justified. This requires rigorous analysis and a compelling business case to prove that the value of the differentiation outweighs the long-term costs of customization.

The third category is critical integration. While SAP offers extensive integration capabilities, there may be instances where connecting to a highly specialized external system (e.g., a proprietary IoT platform, a unique legacy system that cannot be decommissioned, or a niche industry-specific application) requires custom interfaces or data transformation beyond standard adapters. These integrations must be strictly limited to what is essential for business continuity and value.

Even in these justified scenarios, the customization should be meticulously designed, rigorously documented, thoroughly tested, and strategically managed. It should be the extreme exception, the last resort after exploring all standard options, process reengineering, and industry solutions.

Here is a useful guideline to apply for every proposed customization request: If you cannot explain in a single clear sentence how this specific customization drives measurable, strategic business value (e.g., direct revenue, significant cost savings, regulatory compliance, or clear competitive advantage), then you should not do it.

THE ROLE OF CONSULTANTS: CHALLENGER, NOT ENABLER, OF EXCESS CUSTOMIZATION

Let's revisit the role of your consulting partners. A truly good and responsible SAP consultant doesn't just passively take orders from business users. When a user or department says, "We need SAP to work exactly like this legacy system did," a responsible consultant should immediately respond with a probing question: "Why?"

Sometimes, that "why" will reveal a truly legitimate business need that can, in fact, be met by standard SAP functionality in a slightly different, often more efficient, way. The consultant's role is to educate and guide the user to that standard process.

Other times, the "why" exposes an outdated habit, a historical workaround, a siloed process, or a preference that offers no strategic value. In these cases, the consultant's role is to challenge the request, highlight the long-term costs of customization, and advocate for process reengineering or standard adoption.

Unfortunately, and it must be stated frankly, some consulting firms or individual consultants may inadvertently or deliberately enable excessive customization. Why? Because extensive customization translates directly into more billable hours, larger project budgets, and a perception of meeting every client demand. This short-term financial gain for the consulting firm comes at the immense long-term expense of the client. This dynamic underscores why leadership must remain hypervigilant and demand strategic justification for every customization request.

STRATEGIES TO AVOID THE TRAP: BUILDING FOR AGILITY

How can you, as an executive leader, effectively steer your organization away from the perilous customization trap and build an SAP system that is lean, agile, and future-proof? Here are some proven, actionable strategies:

- Adopt a cloud mindset (even for on-premise): In the modern cloud era, less customization is no longer just a "best practice"; it's increasingly a prerequisite for survival and innovation. Cloud-based SAP solutions, particularly SAP S/4HANA Public Cloud, inherently have limited customization capabilities by design, favoring configuration and standard extensions. Treat this architectural constraint not as a burden but as a blessing. Cultivate this "cloud mindset" even if you are implementing an on-premise version of SAP. It forces discipline, promotes standardization, and prepares your organization for a cloud-first future.

- Establish a robust customization governance board: Create a cross-functional governance board or council, comprising senior business leaders (from Finance, Operations, Sales, etc.) and IT leadership. Every single customization request—no matter how small—must be rigorously reviewed and formally approved by this board. The board should ask probing questions: "Does this align with our strategic North Star?" "Is it absolutely necessary for regulatory compliance or competitive differentiation?" "What is the estimated lifetime TCO of this customization (including development, testing, and future upgrade costs)?" "Can this outcome be achieved with a standard process or configuration?" This centralized, disciplined review prevents uncontrolled scope creep.

- Embrace SAP's extension capabilities (SAP Business Technology Platform—BTP): Instead of modifying SAP's core code, leverage SAP's modern extension frameworks, primarily the SAP Business Technology Platform (BTP). BTP allows you to build "side-by-side extensions"—custom applications, workflows, or integrations that run alongside your core SAP system rather than inside it. This approach preserves the integrity of your core SAP system, making it easier to upgrade, while still allowing you to build unique, innovative functionalities that truly differentiate your business. This is the modern, smart way to customize.

- Educate business users proactively and continuously: A significant portion of customization requests stems from a lack of understanding of SAP's standard capabilities. Proactively educate business users early in the project life cycle about how SAP standard processes work. Conduct frequent demonstrations and workshops. Show them that SAP's way is not just different but often superior and more integrated. This helps users understand the "why" behind standard processes and often reveals that their perceived "gap" can be addressed through standard configuration or a minor process adjustment.

- Measure and publicize the long-term ROI/TCO of customization: Develop clear methodologies to calculate not just the up-front cost to build a customization but its projected TCO over five, ten, or fifteen years. This includes maintenance, support, and future upgrade impacts. Publicize these costs transparently within the organization, especially to decision-makers. Quantifying the long-term financial burden often acts as a powerful deterrent to unnecessary customization.

- Challenge legacy thinking and promote process reengineering: Foster a culture where the question "Why do we do it this way?" is welcomed and encouraged. Empower teams to rethink and reengineer their processes to align with SAP's integrated framework. This requires leaders to champion change, reward adaptation, and actively break down departmental silos that often drive unique, inefficient processes.

CASE STUDY: THE MANUFACTURER THAT SAID NO—THE POWER OF DISCIPLINE

Contrast the cautionary tale of the bank with another client, a midsize industrial equipment manufacturer. Early in their SAP implementation, their business users, much like the bank's, submitted dozens of customization requests for everything from unique reporting formats to bespoke inventory management workflows.

Instead of passively approving them, the leadership, guided by a clear vision of standardization, established a dedicated "Customization Council." Every single customization request had to be formally presented to this council with a robust business case, detailing its strategic necessity, why standard SAP could not achieve the outcome, and a realistic estimate of its lifetime TCO. The council, empowered to say no, rigorously scrutinized each proposal. Most requests, lacking a compelling strategic justification or a clear ROI, did not survive this scrutiny.

In the end, this disciplined company went live with only five highly strategic custom developments—all tied directly to nonnegotiable regulatory compliance requirements or critical competitive differentiators. The result? Their SAP system remained lean, clean, and highly integrated. Their upgrade path to SAP S/4HANA Public Cloud was remarkably smooth and fast, achieved in record time, precisely because they had minimal technical debt. Their system was easier to maintain, easier to support, and more adaptable to future innovations.

Saying no to immediate comfort was incredibly hard, generating initial internal friction. But it paid off exponentially in long-term agility, cost efficiency, and strategic enablement.

THE EMOTIONAL SIDE OF CUSTOMIZATION: EMPATHY AND EDUCATION

It's crucial to acknowledge that the drive for customization is often not purely logical; it's deeply emotional. For many employees, their way of working—the specific steps, the unique spreadsheets, the familiar reports—becomes intertwined with their professional identity and expertise. When SAP suggests a different way of working, it can feel like a direct attack on their competence, their value, or their understanding of the business. This emotional component is why simply issuing a mandate to "standardize" often fails.

Leaders must handle this with empathy, education, and genuine engagement. Instead of simply saying, "No, we can't customize that," adopt a more collaborative and educational approach: "Let's explore if SAP already supports what you need in a better, more efficient, and integrated way." Facilitate hands-on demonstrations of the standard process in action. Show them the end-to-end benefits. Often, when users truly understand the integrated power and long-term advantages of SAP's standard processes, they discover that SAP's way is not only workable but indeed superior, empowering them in new ways. It's about managing change through understanding, not just mandate.

In the final analysis, the customization trap is fundamentally about courage. It takes immense courage for leaders to say no to immediate comfort, to resist the easy path of replicating the past, and to challenge the illusion that you can achieve profound business transformation without embracing significant internal change.

Customization promises to preserve the familiar, to keep things "just like before," but in doing so, it often chains your organization to the inefficiencies, costs, and rigidities of the past. Standardization, on the other hand, while requiring initial discipline and adaptation, deliberately positions you for the agility, cost efficiency, and innovation of the future.

The organizations that truly thrive with SAP are those that possess the foresight and discipline to resist the customization trap, courageously embrace standard processes wherever possible, and strategically customize only where it genuinely creates unique, measurable business value.

Every customization represents a critical fork in the road: Will you use SAP as a dynamic, integrated platform for transformative change, enabling new levels of performance? Or will you unwittingly turn it into a costly, cumbersome, and increasingly irrelevant replica of yesterday's habits?

Choose wisely. The future of your business depends on it.

Simplicity Scales, Complexity Kills

When organizations embark on the ambitious journey of an SAP implementation, there's an almost universal and deeply seductive dream: to build the perfect system. This aspiration often involves meticulously capturing every nuance of the existing business, accounting for every conceivable exception, every historical work-around, and every subtle variation in process across departments, regions, or product lines. The vision is to create an ERP solution so comprehensive, so precisely tailored, that it seamlessly accommodates every single operational reality.

It's a noble ambition, born from a desire for completeness and operational precision. Yet, almost invariably, this quest for exhaustive completeness backfires spectacularly. What begins as a seemingly virtuous pursuit of perfection quickly morphs into a convoluted labyrinth of intricate configurations, cascading workflows, bespoke reports, and an unwieldy thicket of exceptions. This sprawling complexity becomes a system that few can genuinely understand, fewer still can effectively use, and even fewer can maintain or evolve without immense effort.

This is the profound paradox of complexity in SAP: The more intricately layered, exception ridden, and functionally bloated your system becomes, the less effective, adaptable, and valuable it ultimately proves to be. While it might appear sophisticated, true power and sustainable growth come not from adding layers of intricacy but from achieving elegant simplicity. Simplicity, not complexity, is what truly scales.

Unlike glaring budget overruns, which scream for immediate attention, or missed go-live deadlines, which create public embarrassment, complexity doesn't announce its arrival with a loud bang. Instead, it creeps in subtly, insidiously, often disguised as perfectly rational, isolated decisions. It's the proverbial frog in slowly boiling water—you don't notice the danger until it's too late.

Consider how complexity quietly infiltrates an SAP project. It often begins with what seems like a small change—"just one extra approval step here." A manager insists on an additional layer of sign-off for a specific type of purchase order, arguing for tighter control. What goes unconsidered is how this slows down the hundreds of thousands of standard purchase orders that must navigate this extra step.

Then there's the demand for "a special case handling there." A business unit demands a unique process flow for a specific customer or product line, believing it provides a competitive advantage. In reality, it creates a costly, hard-to-maintain deviation from the core process, all for a rare scenario that delivers little measurable value.

Complexity also creeps in through unnecessary reporting. An analyst requests a "just in case" report, designed for a hypothetical future scenario. The organization invests time and money into custom development and data storage, only for the report to sit untouched in the system.

Another common source of complexity is the accommodation of legacy work-arounds. Instead of challenging and reengineering outdated, inefficient legacy processes, the project team simply tries to replicate them verbatim in SAP, bringing forward years of accumulated organizational baggage.

Finally, fear of saying no often fuels unnecessary complexity. Project leaders, eager to please stakeholders and avoid conflict, approve requests for unique functionalities without rigorously challenging their strategic necessity or long-term cost.

Individually, each of these decisions might seem small, justifiable, and harmless. Collectively, however, they weave together to create a monstrously complex system—a "complexity debt" that is as crippling as technical debt but often less visible.

The symptoms of this creeping complexity are painfully obvious once you know what to look for.

Onboarding new hires into the SAP environment takes months, not weeks, because the system's logic is convoluted, and exceptions abound. Users and managers constantly argue over the "right" way to execute a transaction, because multiple, often inconsistent, paths exist or the standard process is buried under layers of exceptions. Processes grind to a halt when a seemingly minor exception arises: The system is designed for a multitude of edge cases, but when a truly new or unforeseen scenario occurs, the convoluted logic breaks down, requiring manual intervention and delaying critical operations.

Employees often bypass the new, expensive SAP system for critical tasks because it feels too rigid, too complex, or too difficult to navigate for their daily work. This leads to data inconsistencies and a fragmented operational picture, undermining the very purpose of an integrated ERP. Every upgrade, every new feature, every patch release from SAP becomes a nightmare of testing and rework, ensuring that customized complex layers don't break. This stifles the ability to adopt new, valuable functionalities. Additionally, institutional knowledge is often concentrated in just a handful of highly specialized internal experts or expensive external consultants who truly understand the intricate web of custom processes and configurations, creating a single point of failure and making the organization vulnerable.

If you've seen any of these pervasive symptoms within your organization, complexity is already silently undermining your SAP project, draining value, and inhibiting your potential.

WHY SIMPLICITY WINS: THE POWER OF CLARITY AND FOCUS

Simplicity in SAP design is often misunderstood as cutting corners or sacrificing functionality. On the contrary, simplicity is the ultimate form of sophistication and clarity. A simple, lean SAP design is inherently easier to understand, more straightforward to adopt, and far more capable of scaling with your business's evolving needs.

Here's why simplicity consistently wins in the long run:

- Accelerated adoption and user empowerment: People actually use systems that feel intuitive, straightforward, and logical. A simple user interface, combined with streamlined processes, reduces the learning curve, minimizes frustration, and fosters a sense of empowerment among employees. When users trust and understand the system, they embrace it, maximizing its value.

- Enhanced speed and efficiency: Processes designed with simplicity at their core run faster, with fewer unnecessary steps, fewer handoffs between departments, and fewer points of potential error. This translates directly into faster financial closes, quicker order-to-cash cycles, more agile supply chains, and ultimately, a more responsive business. Think of a streamlined highway versus a tangled web of back roads.

- Greater agility and adaptability: A simple, standardized design is inherently more flexible. When the business needs to change— whether launching a new product, entering a new market, acquiring another company, or responding to regulatory shifts—a lean SAP system can adapt far more easily and cost effectively. Complex systems, burdened by intricate interdependencies and custom code, become rigid and brittle, resisting change and forcing expensive, time-consuming overhauls.

- Increased resilience and stability: Simplicity inherently reduces the number of potential failure points. With fewer custom configurations, fewer bespoke integrations, and less complex process flows, the system is naturally more stable, less prone to bugs, and easier to troubleshoot when issues do arise. This leads to higher uptime, less operational disruption, and greater reliability.

- Lower TCO: A simple SAP system is significantly cheaper to implement (less development, less testing), cheaper to maintain (easier to support, less rework with upgrades), and cheaper to evolve (easier to adopt new features). These long-term cost savings are a direct result of up-front design discipline.

Consider global technology giants like Amazon or Apple. Their customer-facing products and services appear deceptively simple and intuitive. But that simplicity is not accidental; it is the direct result of incredibly disciplined, intentional, and often ruthless design decisions behind the scenes. SAP implementations need to adopt this same unwavering discipline.

CASE STUDY: THE INSURANCE GIANT THAT COLLAPSED UNDER ITS OWN WEIGHT

Let me share a compelling, cautionary tale. I once consulted with a major international insurance giant embarking on an SAP claims management implementation. Their stated goal was to create a system that could handle every conceivable permutation and exception within their complex claims processing. They wanted to capture every possible scenario, every unique regulatory requirement across various jurisdictions, and every historical work-around used by individual claims adjusters.

By the time of go-live, the SAP system had ballooned into a Frankenstein of complexity. It boasted over eight hundred unique workflows for claims processing, each with its own set of rules, fields, and approval paths. The intention was to provide ultimate flexibility, but the reality was chaos.

The results were predictable:

- User paralysis: Claims adjusters were utterly paralyzed. Nobody could remember which of the eight hundred workflows applied to a specific claim. Simple tasks became complex decision trees, leading to indecision and errors.

- Training nightmare: Training new employees on the system took many months, and even experienced staff constantly struggled. The sheer volume of exceptions made it impossible to build intuition or muscle memory.

- Flourishing work-arounds: Frustrated users quickly developed extensive manual work-arounds outside the system—spreadsheets, email chains, phone calls—to handle claims, completely undermining the purpose of an integrated ERP.

- Massive cost of errors: The complexity led to a significant increase in processing errors, costing the company millions in reprocessing fees, regulatory fines, and customer dissatisfaction.

Eventually, after years of struggle and mounting operational costs, leadership painfully realized they had overbuilt the system. They had to launch a second multimillion-dollar project specifically focused on simplifying the very SAP system they had just implemented. This involved decommissioning hundreds of workflows, standardizing processes, and accepting that some rare exceptions would need manual intervention. It was a costly and humiliating lesson: Their quest for comprehensive perfection had effectively collapsed their core business operations under its own weight.

THE ILLUSION OF "FUTURE-PROOFING": A MYOPIC TRAP

One of the most insidious justifications for adding complexity to an SAP system is the belief that leaders are "building for the future." This often translates into designing elaborate processes, adding countless optional fields, or building functionalities for hypothetical markets or products that may never materialize.

The logic seems sound: Prepare for every contingency, build for every potential growth path, and ensure the system can handle anything. However, this often leads to a dangerous form of analysis paralysis and overengineering.

I've seen companies invest heavily in designing intricate processes for markets they never entered or adding layers of approval for risks that never materialized. They obsessed over minute details for scenarios that had a 1 percent chance of occurring, while neglecting to optimize the 99 percent of daily operations. They believed they were "future-proofing" themselves, but in reality, they were building a rigid, overly complex system that future-proofed them into paralysis.

The truth is, the best way to truly future-proof your SAP system is to keep it simple, agile, and adaptable. The business landscape will always change in unpredictable ways. New technologies will emerge, customer demands will shift, and competitive dynamics will evolve. A lean, simple system, free from excessive customization and complexity, can readily adapt to these unforeseen changes. A sprawling, complex one cannot. It will resist change, making every adaptation a costly, time-consuming ordeal.

Moreover, building for an imagined future often leads to "shelfware"—features, functionalities, or entire modules that are implemented but never actually used. This represents millions of dollars in wasted licenses, implementation costs, and ongoing maintenance for capabilities that provide zero business value.

THE 80/20 RULE OF PROCESS DESIGN: OPTIMIZING FOR THE CORE

A fundamental principle that should guide every SAP design decision is the 80/20 rule (Pareto principle): Design your core processes to efficiently handle the 80 percent of transactions or scenarios that are common and frequent rather than overengineering for the 20 percent of rare exceptions.

In practice, applying this principle to SAP design means focusing on the following key actions:

- Standardize the common scenarios: Identify the high-volume, repetitive processes (e.g., standard sales orders, basic procure-to-pay for regular suppliers, routine financial postings). These are the processes where standardization, automation, and simplicity yield the greatest returns. Design SAP to handle these with minimal clicks, clear flows, and integrated data.

- Streamline the main process: Focus on creating the leanest possible path for the majority of your business transactions. Eliminate unnecessary steps, redundant approvals, and manual handoffs that don't add value.

- Handle exceptions manually if needed, or with simple extensions: Don't let rare or complex scenarios dictate the entire system design for everyone. If a specific type of order, a unique return process, or a niche regulatory report occurs only a few times a year, it might be more cost effective and simpler to handle it manually (perhaps with a simple, isolated BTP extension) than to build complex, rarely used logic into the core SAP system that complicates everyday operations for everyone else.

- Don't overengineer for the edge cases: If 80 percent of your customer orders follow the same streamlined path, optimize that path rigorously. Do not add layers of complexity, custom fields, or alternative workflows for the infrequent 20 percent of edge cases. These exceptions can often be managed outside the core flow or with very targeted, isolated solutions.

Adhering to the 80/20 rule allows the SAP system to operate with maximum efficiency for the majority of your business, making it faster, easier to use, and more stable.

HOW COMPLEXITY CREEPS IN: THE BEST INTENTIONS GONE AWRY

Complexity rarely arises from malice or deliberate sabotage. More often, it stems from seemingly good intentions, compounded by a lack of disciplined oversight and strategic clarity. Individual business units advocate for their unique requirements, convinced that their processes are distinct and indispensable. Without a strong, unified vision from the top and a clear "best fit" strategy, every departmental "special need" can become a customization, leading to a fragmented, overengineered system. Leaders, wanting to cover every possible risk or future scenario, may introduce overly complex controls, redundant approval workflows, and excessive data capture, adding layers of bureaucracy that slow down operations without proportional benefit, which often stems from a fear of future audit findings or unforeseen operational issues.

Consultants, too, may over-deliver to impress clients or ensure no requirement is missed, proposing elaborate solutions or extensive customization to demonstrate their technical prowess or to ensure no requirement is missed. This can lead to building functionalities that are "nice to have" but not strategically essential, adding unnecessary complexity. When no single individual or body is empowered to make difficult trade-offs and say no to complexity, decisions are often deferred or compromised or fall victim to political maneuvering, leading to the accumulation of features that serve no coherent purpose.

Finally, the sheer volume of options within SAP can overwhelm teams, leading to endless debates and an inability to make clear, simple design choices. This indecision often results in "compromise solutions" that are overly complex and satisfy no one optimally.

The danger is that nobody feels individually responsible for the overall complexity. Everyone adds "just one more thing" until the system collapses under the collective weight of these seemingly small, individual decisions.

TOOLS TO FIGHT COMPLEXITY: DISCIPLINED STRATEGIES FOR SIMPLICITY

So how do you actively guard against the silent killer of complexity and ensure your SAP system remains lean and effective? It requires deliberate, proactive strategies and unwavering discipline:

- Adopt standard SAP first—the default position: Start every design discussion with the question "How does standard SAP address this business requirement?" Only after exhaustively exploring and understanding the standard options should you consider deviations. Customization should always require a compelling, strategically justified business case and be seen as an exception, not the rule.

- Challenge every exception and deviation: For every request that deviates from standard SAP or adds a layer of complexity, rigorously ask, "Is this truly necessary? What specific business value does it provide? What is the lifetime cost and impact on simplicity? Can this outcome be achieved through process reengineering or manual handling for rare cases?" Demand quantitative justification.

- Map processes visually (and keep them simple): Utilize tools like value stream mapping to visually represent your "to be" processes. If a process flow diagram doesn't fit on a single page or requires an excessive number of swim lanes and decision points, it's a strong indicator that it's already too complex. Aim for lean, intuitive flows that eliminate bottlenecks and unnecessary handoffs.

- Create a "simplicity champion" or design authority: Appoint a dedicated internal leader or establish a specific governance body (e.g., a design authority board) whose explicit role is to actively push back against overdesign, challenge complexity, and champion standardization. This role requires courage and a deep understanding of both business strategy and SAP capabilities. They are the guardians of simplicity.

- Review metrics regularly for complexity symptoms: Track key performance indicators (KPIs) that can reveal hidden complexity. This includes

  - Process cycle times: How long does it take for an order to cash or a vendor invoice to be paid? Long cycle times often indicate complex, multistep processes.

  - Manual work-arounds: How often are users exporting data to spreadsheets, sending emails for approvals, or performing tasks outside SAP? This is a clear sign of complexity and lack of system fit.

  - User support tickets: A high volume of tickets related to process confusion or how-to questions can indicate overly complex system navigation or illogical workflows.

  - Customization ratios: Track the number of custom reports, interfaces, conversions, enhancements, and forms/workflows (RICEFW objects). Aim to keep this ratio as low as possible.

- Implement a minimum viable product (MVP) approach for go-live: Instead of trying to build the "perfect" system with every conceivable feature for the initial go-live, focus on implementing only the core, essential functionalities that deliver immediate business value. Prioritize the must-haves for go-live, and defer nice-to-haves to later phases. This reduces initial complexity, accelerates time to value, and allows for iterative refinement.

CASE STUDY: THE RETAILER THAT SIMPLIFIED TO SCALE—THE POWER OF DISCIPLINE

Contrast the insurance giant's struggles with a highly successful transformation story from a global retailer. Facing the need to modernize their operations for aggressive international expansion, the insurance company's leadership began their SAP S/4HANA journey with an unwavering commitment to a "simplicity first" policy. This wasn't just a slogan; it was rigorously enforced.

From the project's inception, every single request from every department had to be justified not only by a perceived business need but, crucially, by its impact on the system's overall simplicity. When a procurement department manager requested a highly unique, multilevel approval workflow for a specific category of indirect spend, leadership directly asked, "Will this unique workflow make our overall procure-to-pay process faster, more efficient, and easier for the majority of our users, or will it make it slower and more complex?" The honest answer, after analysis, was slower and more complex for the majority. Therefore, despite the department's strong preference, the request was denied in favor of a standard, streamlined process.

This unwavering discipline, consistently applied at every decision point, led to remarkable results. By go-live, the retailer had a remarkably clean, lean, and highly standardized SAP system. User adoption was exceptionally high because the processes were intuitive and logical. Operational processes ran at unprecedented speeds, reducing cycle times and improving responsiveness. And, most impressively, when the retailer embarked on its ambitious expansion into new international markets, the lean, simple SAP system scaled effortlessly, requiring minimal local adaptation and enabling rapid market entry at a fraction of the cost and time of their competitors. Saying no was hard, and it generated initial internal friction, but it paid off exponentially in long-term agility and global scalability.

Interestingly, many leaders, and indeed many project teams, struggle deeply with the concept of embracing simplicity because, on the surface, it can feel less impressive. A complex, bespoke design, with its myriad features and intricate workflows, can look sophisticated, advanced, and comprehensive. It might even flatter the ego, suggesting a deep understanding of every possible nuance. A truly simple design, by contrast, looks . . . well, simple. It doesn't necessarily scream "genius" to the uninitiated.

But that's precisely the dangerous trap. Complexity flatters the ego and provides a false sense of security, while true simplicity unequivocally serves the business. Real leadership in an SAP project requires the humility to choose simplicity over grandeur, effectiveness over perceived sophistication, and long-term agility over short-term comfort. It means being courageous enough to make the difficult choices to streamline, standardize, and eliminate, even when met with internal resistance.

SAP CLOUD AS A FORCING FUNCTION: A BLESSING IN DISGUISE

One of the hidden blessings and powerful strategic drivers of modern SAP S/4HANA Public Cloud implementations is that it inherently limits complexity by design. In a multi-tenant cloud environment, extensive core customization is simply not allowed or is severely restricted. Customers cannot modify the underlying code as freely as they could in older on-premise SAP ECC systems. Many core processes come highly standardized, and new functionalities are delivered regularly as part of a standardized release cycle.

Some executives initially perceive this as a significant restriction or a frustrating limitation on their business uniqueness. In reality, it is profoundly liberating. This architectural constraint forces organizations to embrace standardization. It demands that companies adapt their processes to SAP's best practices rather than bending SAP to their legacy ways. It also forces organizations to prioritize configuration and extensions. It encourages leveraging standard configuration capabilities and building side-by-side extensions on platforms like SAP Business Technology Platform (BTP) for unique needs rather than modifying the core. Additionally, organizations are forced to focus on agility by keeping the core clean. It significantly reduces technical debt, making the system easier to maintain, upgrade, and adapt to future innovations.

This "forcing function" of the cloud paradigm compels organizations to adopt a simpler, leaner approach to SAP, which, in turn, makes them fundamentally more agile, more cost effective, and ultimately, more competitive. It's about getting out of your own way to leverage the power of a standard platform.

SIMPLICITY AS A COMPETITIVE ADVANTAGE: BEYOND JUST IT COSTS

In the long run, simplicity in your SAP system is far more than just an IT cost-reduction strategy. It directly translates into these tangible, powerful business results that drive competitive advantage:

- Faster onboarding for new employees: Reduced training times mean new hires become productive more quickly, lowering HR costs and accelerating team growth.

- Faster, more accurate decision-making: Simple, clear data models and streamlined reporting processes provide leaders with rapid access to reliable information, enabling quicker and more informed strategic and operational decisions.

- Faster market entry and global expansion: A lean, standardized core system can be replicated and deployed in new geographies or acquired entities with unprecedented speed and minimal friction, accelerating growth strategies.

- Improved customer responsiveness: Streamlined, integrated processes (from order-to-cash or service-request-to-resolution) lead to quicker fulfillment, more accurate information, and ultimately, a superior customer experience that builds loyalty.

- Enhanced employee morale and satisfaction: Users who find the system intuitive and empowering are more engaged, less frustrated, and more productive, leading to a more positive work environment.

- Reduced operational risk: Fewer complexities mean fewer points of failure, more consistent processes, and easier compliance.

Complexity, by stark contrast, becomes a debilitating tax on every single part of the business—a drag on efficiency, innovation, and profitability.

The truth, stark and undeniable, is this: Simplicity in an SAP implementation doesn't happen by accident. Complexity is the pervasive default; it creeps in naturally, silently, often with the best of intentions. Achieving genuine simplicity requires unwavering discipline—to saying no to unnecessary customization, to redundant features, to special requests that don't align with strategic priorities. This also means having the discipline to design lean, to constantly challenge existing processes, to strip away complexity, and to optimize for the 80 percent that drives the most value. Lastly, this is to have the discipline to focus on the vital few instead of getting lost in the trivial many.

In SAP projects, just as in life and in business strategy, it's often tempting to equate complexity with sophistication and simplicity with a lack of depth. But the organizations that truly thrive with SAP are the ones that deeply understand this fundamental principle: Simplicity scales and complexity kills. They choose discipline over passive drift.

By making this choice, they transform their SAP system from a potential source of frustration and cost into a powerful, agile platform that genuinely amplifies their strategy and drives sustainable competitive advantage.

Choosing the Right Partner, Not Just the Cheapest

When organizations commit to an SAP implementation, they are, in essence, making one of the most significant and transformative investments in their corporate history. This decision typically involves multimillion-dollar outlays, profound operational changes, and a reshaping of their future capabilities. Understandably, cost is, and always will be, a significant factor in such a monumental undertaking. However, far too often, executive leadership mistakenly treats the partner selection process as if it were a routine procurement exercise, akin to buying office furniture or commodity software licenses: Solicit three competitive quotes, mechanically choose the lowest bidder, and then swiftly move on.

This simplistic, cost-driven mindset is not just problematic; it is, quite frankly, disastrous when it comes to SAP.

Your SAP implementation partner is fundamentally not just another vendor; they are, in the truest sense, your copilot on a complex, high-stakes journey that will profoundly reshape how your entire business operates for the next decade, if not longer. This is a journey fraught with technical challenges, operational complexities, and significant organizational change. The consequences of choosing the wrong copilot can be catastrophic: The project can derail entirely, invaluable internal resources will be drained, costs will spiral out of control, and you could be left with a new SAP system that is either technically dysfunctional or, perhaps worse, one that nobody within your organization genuinely trusts or wants to use.

Conversely, the right partner can become an indispensable asset. They can expertly guide you through the inherent complexity, anticipate pitfalls, and courageously challenge your assumptions when necessary. They will transfer knowledge to your internal teams, empower your people, and ultimately leave your organization stronger, more agile, and strategically better positioned for the future. This is not merely a procurement decision; it is, at its core, a strategic transformation decision that demands far greater scrutiny, foresight, and executive involvement than a typical vendor selection process.

On paper, at the outset of the project, choosing the cheapest implementation partner might appear to be a financially prudent decision—a clear path to significant budget savings. In the short term, the initial proposal from a low-cost provider might indeed present a more attractive numerical figure. However, in the vast majority of cases, these seemingly advantageous "savings" are nothing more than a dangerous illusion. In reality, opting for the cheapest partner almost invariably ends up costing the organization far more in the long run, often exponentially so.

This hidden cost arises because "cheap" partners typically achieve their lower price points by systematically cutting corners in precisely the areas that matter most for a successful, sustainable SAP implementation. Often, they rely on inadequate staffing and junior consultants. Low-cost providers often staff projects with a higher proportion of junior, less experienced consultants who lack the deep industry knowledge, the nuanced SAP expertise, or the critical problem-solving skills necessary to navigate complex business challenges. While cheaper, their lack of foresight can lead to repeated mistakes, extensive rework, and missed opportunities for process optimization.

Another common shortcut is rushing through critical phases. Project timelines are compressed unrealistically, particularly for crucial phases like design, blueprinting, and testing. This leads to superficial design sessions where the partner fails to truly understand your unique business requirements, overlooks critical interdependencies, or simply imposes generic "best practices" without proper contextualization. This leads to misaligned configurations and a system that doesn't genuinely fit your needs.

Substandard deliverables are another hallmark. To meet aggressive deadlines and stay within budget, cheap partners often produce half-baked documentation, incomplete process flows, or insufficient training materials. This leaves your internal team scrambling after go-live, struggling to understand and maintain a system that lacks proper instructional support.

Compounding this problem is the neglect of knowledge transfer and change management. A critical role of a good partner is building internal capability and preparing your team for self-sufficiency post-go-live. Cheaper firms often deprioritize robust knowledge transfer sessions, comprehensive training, and proactive change management activities, leaving your organization ill-equipped to operate, support, and evolve the new system independently.

Finally, poor risk management is a frequent weakness. Low-cost bids often fail to adequately account for inherent project risks (data quality issues, integration complexities, scope changes). When these inevitable risks materialize, they lead to unforeseen costs, delays, and frustrating finger-pointing, ultimately escalating the total project expenditure.

The superficial, short-term "savings" from choosing the cheapest partner quickly evaporate—and often turn into a significant financial deficit—when your organization is inevitably forced to fix what wasn't done right the first time. This might involve reengaging consultants to correct fundamental design flaws, hiring additional internal staff to manage ongoing support issues, or delaying the realization of promised business benefits.

There's an old, painfully true saying in the world of complex projects: "If you think good consultants are expensive, wait until you hire bad ones." The cost of remediation, lost opportunity, and diminished ROI from a poor partner dwarfs any initial savings.

CASE STUDY: THE MANUFACTURER WHO PAID TWICE—A PAINFUL LESSON IN VALUE

Consider the experience of a midsize industrial manufacturer. Driven by a compelling directive to reduce initial project expenditure, their leadership team ultimately chose the lowest-bid implementation partner from a field of several proposals. The chosen partner promised an aggressive timeline and a price point significantly below competitors.

However, the "savings" quickly unraveled. The partner primarily staffed the project with junior, less experienced offshore consultants who, despite being technically proficient, lacked the deep industry understanding specific to the client's complex manufacturing processes. This meant that critical business nuances were missed during design sessions, leading to fundamental configuration flaws that didn't align with the manufacturer's operational reality. Data migration, already a complex undertaking, became an unmitigated nightmare due to inadequate planning and a lack of experienced data specialists. The project was riddled with constant rework and escalating internal frustration.

Six months after a tumultuous go-live, the SAP system was technically operational but functionally crippled. Processes were breaking down regularly, financial reporting was inconsistent, and users were resorting to extensive manual work-arounds outside the system. The manufacturer's leadership was faced with a painful truth: They had to bring in a second, higher-quality, and more expensive consulting partner to "rescue" the system—to fix the fundamental design flaws, clean up the data, and stabilize operations. The final, cumulative cost of the initial "cheap" implementation plus the subsequent "rescue" effort ended up being more than double what the higher-quality, initially more expensive partners had originally quoted.

This was a devastating lesson for the manufacturer: The cheapest option, pursued with a short-term financial focus, had turned out to be the most expensive and damaging path in the end. They had effectively paid twice for their SAP system, while enduring years of operational pain and delayed benefits.

WHAT TO LOOK FOR IN A PARTNER: BEYOND THE PRICE TAG

So, if cost is not the primary determinant, what are the essential attributes and characteristics you should rigorously evaluate when selecting an SAP implementation partner? Focus on these critical factors, understanding that they directly correlate with long-term success and value:

- Industry expertise and domain knowledge: Does the partner genuinely understand your specific industry sector? Have they successfully implemented SAP for companies with similar business models, regulatory environments, and competitive landscapes? This is paramount. A partner with deep industry expertise will speak your language, grasp your unique challenges without extensive explanation, anticipate specific requirements (e.g., batch management in pharma, project systems in engineering, trade promotions in consumer goods), and offer pre-built industry templates or accelerators that are actually relevant to your business. They will bring "been there, done that" wisdom that prevents costly missteps.

- Proven methodology and project management rigor: Do they have a clear, structured, and repeatable approach to SAP implementations? A strong methodology (e.g., leveraging SAP Activate principles with agile iterations) provides a road map, defines phases, and outlines key deliverables. Beyond the methodology, assess their project management rigor:

  - How do they handle scope management? Do they have a robust change control process?

  - What is their approach to risk identification and mitigation?

  - How do they manage quality assurance, testing, and defect resolution?

  - What are their strategies for knowledge transfer and building internal capability within your team?

  - How do they integrate change management into every phase, not just as an afterthought? A mature methodology indicates discipline and predictability, reducing chaos.

- Quality of people (not just quantity): This is perhaps the single most important factor. Ask explicitly: "Who will actually be on our project team?" Demand to interview the proposed project manager, solution architects, and key functional leads. Assess their experience not just with SAP but with similar business challenges, their problem-solving approach, and their ability to communicate complex concepts clearly. A core team of senior, experienced experts, even if smaller, is infinitely more valuable than a large bench of junior, inexperienced consultants. A team with a track record of collaboration and successful delivery is essential.

- Cultural fit and communication style: An SAP implementation is a profoundly collaborative effort. Does the partner's team demonstrate a cultural fit with your organization? Do they communicate openly, honestly, and respectfully? Can they challenge your assumptions constructively without being antagonistic? Do they prioritize active listening and understanding over simply selling a prepackaged solution? A good cultural fit fosters trust, open dialogue, and a genuine partnership, which is vital for navigating difficult discussions and making tough decisions.

- Verifiable track record and references: Don't just rely on polished marketing slides or generic client lists. Ask for and diligently contact real customer references—ideally, companies of similar size, in similar industries, who have recently completed projects with this partner. Ask references about specific challenges, how the partner handled them, their commitment to post-go-live support, and whether they would reengage the same partner. Also, research the partner's financial stability and long-term commitment to the SAP ecosystem.

- Focus on value, not just cost: Does the partner's proposal clearly articulate the value they bring, beyond just their day rate? Do they demonstrate how their approach will help you achieve your strategic business outcomes (e.g., faster financial close, reduced inventory, improved customer experience), not just a technical go-live? A focus on value indicates a true partnership whereas an obsession with being the cheapest often signals a transactional, short-term approach.

Notice what is conspicuously missing from this list of critical factors: the raw day rate or the total proposed project price as the primary determinant. While price is a consideration, it should never overshadow these qualitative measures of capability, expertise, and partnership.

A PARTNER IS A MIRROR: THE VALUE OF HONEST DISAGREEMENT

A truly great SAP partner does more than just configure software or provide technical resources; they act as a vital mirror to your organization, reflecting back insights, challenges, and uncomfortable truths that you might otherwise overlook or avoid confronting. They will have the courage and integrity to tell you when your existing business processes are fundamentally broken and need reengineering, not just replication in SAP. They will tell you when your master data is not nearly as ready as you believe it is, and will cause significant issues if not addressed. They will communicate when your executive leadership team is not truly aligned on the project's vision or key strategic decisions, creating internal conflict that will derail the implementation.

Additionally, they will discuss a particular customization request, which, while desired by a department, will introduce unacceptable long-term costs and complexity and does not serve the overall business strategy.

Bad partners, by contrast, avoid conflict at all costs. They are yes-men consultants. They will say yes to every demand, implement what you ask for without challenge, and avoid delivering uncomfortable truths, even when it is clearly not in your best long-term interest. They prioritize their own billable hours and avoid client friction over ensuring your project's long-term success.

You do not need a partner who simply agrees with everything you say; you need a partner who possesses the strategic acumen and integrity to disagree, respectfully but firmly, when it truly matters—a partner who cares enough about your success to challenge you, educate you, and guide you toward the optimal path. This candid feedback, while sometimes difficult to hear, is incredibly valuable in navigating the complexities of an SAP transformation.

RED FLAGS TO AVOID: WARNING SIGNS OF A TROUBLED PARTNERSHIP

When you are engaging with potential SAP implementation partners, be acutely aware of certain red flags that should trigger immediate caution and deeper scrutiny. These warning signs often indicate a troubled partnership waiting to happen. One of the most obvious red flags is aggressive over-promising. Be highly skeptical of partners who claim things like "We can implement SAP in twelve weeks, no matter your scope or complexity!" or "We guarantee X percent cost savings from day one!" Unrealistically aggressive timelines or unquantified guarantees often indicate a lack of understanding of your business, a willingness to cut corners, or an attempt to win a bid by making impossible promises.

Another common issue is the bait-and-switch approach to staffing, a common tactic where highly experienced senior consultants are prominently featured during the sales cycle and in initial proposals, only for junior, less experienced (and cheaper) resources to appear on the actual project team once the contract is signed. Demand to meet and approve the actual core project team members before signing.

Lack of transparency is also a major red flag. If a partner offers vague or evasive answers about their specific methodology, the composition of their project team (e.g., "we'll tell you later"), their approach to risk management, or their post-go-live support model, it's a clear warning sign. A reputable partner will be open and direct about these critical aspects.

An exclusive focus on price should also give you pause. If the partner's primary conversation revolves around their lower cost, aggressive discounts, or hourly rates rather than a detailed discussion about the value they will deliver, their proposed solution, their industry experience, or their team's quality, it's a major red flag. This indicates they are competing on price, not value.

Similarly, boilerplate proposals and a lack of customization can indicate a lack of real commitment to your success. If their proposal feels generic, recycling content from other bids, and doesn't demonstrate a deep understanding of your unique business challenges, objectives, or industry nuances, it suggests they haven't invested the time to truly understand your needs.

References—or lack thereof—can be telling as well. If they provide only a limited number of references, or if the provided references are difficult to contact or reluctant to speak, it could be a warning sign. A strong partner will have a long list of happy clients willing to vouch for them.

Finally, beware of partners who cannot explain mistakes or challenges. No project is perfect. A good partner can openly discuss projects where they faced significant challenges, what mistakes they made, and, crucially, how they learned from those experiences and improved their approach. A partner who claims a flawless record is either dishonest or inexperienced.

If you encounter these red flags during your partner selection process, heed the warning signs. Walking away from a problematic partnership up front is far less costly and damaging than enduring a disastrous one.

An SAP implementation is rarely a "one and done" project; it's the beginning of a long-term strategic relationship with the software and, typically, with your implementation partner. After the initial go-live, your organization will inevitably require ongoing support, continuous enhancements to leverage new SAP functionalities, and potentially new module rollouts or global template deployments. Your chosen partner will likely be an integral part of your ecosystem for years, perhaps even a decade or more.

Think about this relationship not as a transactional vendor engagement but as a strategic, long-term partnership. Would you "marry" someone based solely on their hourly wage or the lowest price point? Of course not. You would choose a partner with whom you can build a future—someone you trust, who shares your values, who communicates effectively, and who is committed to your long-term success. Partner selection for SAP deserves precisely the same seriousness, diligence, and foresight as any other critical long-term strategic alliance your company might forge. The partner's ability to evolve with your business, understand your changing needs, and provide ongoing strategic advice is paramount.

CASE STUDY: THE GLOBAL ROLLOUT THAT WORKED—INVESTING IN PARTNERSHIP

Consider the transformative experience of a large global manufacturer of industrial equipment. They embarked on an ambitious, multiphase SAP S/4HANA rollout across twelve distinct countries and business units. Rather than selecting the cheapest bidder, their leadership made a deliberate, strategic decision to choose a partner not for their low price but for their undeniable deep expertise in the complex industrial manufacturing sector and their proven track record with large-scale global template deployments.

Yes, this chosen partner's initial proposal was more expensive up front. However, their superior methodology, highly experienced team, and proactive approach to change management proved invaluable. The partner meticulously guided the manufacturer through each phased rollout, anticipating and mitigating risks, expertly handling localizations, and ensuring consistent user adoption across diverse cultures. They acted as true strategic advisors, challenging the client when necessary and celebrating successes as a shared team.

The result was exceptional: The SAP system worked flawlessly, adoption rates were remarkably high across all regions, and the manufacturer was able to consolidate operations, gain real-time global visibility, and scale efficiently into new markets with unprecedented speed. Over the next ten years, the initial "more expensive" investment paid for itself many times over in terms of operational efficiency, strategic agility, and quantifiable business benefits. The long-term value far, far outweighed the initial cost difference.

QUESTIONS TO ASK IN PARTNER SELECTION: PROBING FOR TRUE VALUE

To truly discern the right partner, you must ask probing questions that go beyond superficial capabilities. Here are questions I strongly encourage clients to ask every potential partner, both during formal request for proposals (RFPs) and in one-on-one executive interviews:

- Team quality and engagement
  - "Who will be the specific individuals on our core project team (project manager, solution architects, key functional leads), and can we interview them extensively before final selection?"
  - "What is their specific experience with our industry and similar business transformations?"
  - "How do you ensure continuity of key resources throughout a multiyear project?"
- Methodology and execution
  - "Beyond your standard methodology, how do you adapt your approach for a client of our size, complexity, and unique strategic goals?"
  - "How do you handle scope changes, conflicts between business units, or unexpected data quality issues during the project?"
  - "What is your philosophy on customization versus standardization, and how do you ensure our strategic priorities drive this decision?"
  - "Describe your approach to comprehensive testing (unit, integration, user acceptance, performance) and training. How do you ensure users are truly ready at go-live?"
  - "How do you plan for knowledge transfer to our internal team to ensure we are self-sufficient post-go-live?"
- Partnership and trust
  - "Give us an example of a time your firm disagreed strongly with a client's request or strategy and how you navigated that situation. What was the outcome?"

- "Tell us about a project you had to 'rescue' after another partner failed. What were the root causes of that failure, and what did you do differently?"

- "What significant mistakes has your firm made on a large SAP project, and what concrete lessons did you learn from them?" (This question reveals humility and a commitment to continuous improvement.)

- "How do you measure project success beyond simply 'going live'?" (Look for answers that tie to business benefits and value realization.)

- Long-term vision

- "How do you envision our partnership evolving beyond the initial go-live, particularly as we consider future SAP innovations or global rollouts?"

The answers to these types of probing questions will reveal far more about a partner's true capabilities, integrity, and cultural fit than any slick marketing presentation or detailed technical proposal ever could.

WHY PROCUREMENT OFTEN GETS IT WRONG: THE MISMATCH OF PARADIGMS

It's a common organizational challenge: Many companies, in their commendable pursuit of financial discipline, delegate the SAP partner selection process almost entirely to their procurement departments. Procurement teams, by their nature and training, excel at running RFPs, establishing detailed scoring matrices, and ultimately awarding contracts to the lowest bidder, particularly for commoditized goods and services.

However, an SAP implementation partner is decidedly not a commodity. It is a complex, strategic, human-intensive partnership that demands deep collaboration, trust, and shared risk. Applying a purely commoditized procurement approach—where the lowest bid wins—almost guarantees disappointment and ultimately leads to higher long-term costs. Procurement's focus on cost minimization, while necessary for some categories, is often diametrically opposed to the need for value maximization, cultural fit, and strategic alignment in an SAP project.

The companies that truly succeed with SAP are almost always those where executive leadership gets personally and deeply involved in the partner selection process. They look far beyond the initial cost figures, prioritizing the value proposition, the quality of the team, the depth of industry expertise, and the long-term partnership potential. They understand that this is a strategic investment in their future operating model, not just a transaction.

Just as simplicity scales in system design, it also applies to the partner selection process itself. Don't fall into the trap of designing a three-hundred-question RFP that overwhelms both your internal team and potential partners, leading to generic, voluminous responses that mask a lack of depth. Instead, focus on the vital few, high-impact criteria that truly differentiate a good partner from a bad one:

- Can they do the work? (expertise, methodology, track record)

- Do they understand us? (industry experience, cultural fit, ability to challenge)

- Do we trust them? (transparency, honesty, references)

Everything else, while potentially useful data, is largely noise compared to these core elements. A disciplined, focused evaluation based on these questions will lead to a far more effective partnership.

At the end of the day, choosing the right SAP implementation partner is, without exaggeration, one of the most important decisions your organization will make in its digital transformation journey. It is tempting, perhaps even pressured, to prioritize up-front savings by defaulting to the cheapest bidder. However, the hard-won lessons from countless projects unequivocally demonstrate that this path often leads to exponentially higher costs, excruciatingly longer timelines, profound operational disruption, and ultimately, a pervasive erosion of internal trust and confidence.

The right partner may indeed carry a higher up-front price tag. But their investment in quality, their depth of expertise, their commitment to your success, and their ability to act as a true strategic copilot will save you far more in the long run—not just in direct dollars and cents but in the invaluable currency of successful business outcomes, elevated team morale, accelerated strategic agility, and the enduring competitive advantage that a well-implemented SAP system delivers.

Because when it comes to an SAP transformation, the stark reality is this: Cheap is almost always expensive, and the right partner is, quite literally, priceless. Their guidance will be the difference between merely going live and truly transforming.

Driving Real Value from Your SAP Project

If you've ever had the experience of leading or being involved in a major business initiative, particularly a large-scale technology implementation, you've likely witnessed a deeply frustrating and all-too-common pattern: The project team diligently delivers every single item on the meticulously crafted project plan. Deadlines are met, budgets are adhered to, countless documents are handed over, and the new systems are, technically, successfully switched on. And yet, despite all this apparent "success," the business walks away with a palpable sense of disappointment. Boxes were checked, tasks were completed, and systems are live—and yet the profound, transformative change that leaders earnestly expected never quite materializes.

This unsettling disconnect lies at the heart of the trap of focusing solely on deliverables. Deliverables are, without question, necessary components of any well-managed project. They represent the building blocks, the tasks completed, and the tangible outputs. But here's the critical distinction: Deliverables are not sufficient on their own. What genuinely matters, what ultimately defines true success in an SAP implementation, is whether those completed deliverables genuinely translate into measurable outcomes—tangible, positive improvements in your business. We're talking about real gains in operational efficiency, a noticeable enhancement in customer experience, more insightful decision-making, improved profitability, and sustained long-term organizational agility.

SAP implementations are particularly notorious for this disconnect. Organizations spend hundreds of millions of dollars, receive stacks of meticulously detailed functional specifications, execute thousands of test cases, and proudly announce "successfully" configured modules. But when leaders, months after go-live, look around and ask, "Has our business truly transformed? Are we solving the problems we set out to solve? Are we realizing the value we expected?", the response is often an awkward silence, a shrug of the shoulders, or a defensive enumeration of technical achievements rather than business impacts.

This is a crucial and fundamental shift in mindset: moving your focus from simply measuring your SAP project by what got delivered to rigorously measuring it by what problems it actually solved, what value it created, and what outcomes it achieved.

Many executives and project managers, when approaching an SAP project, instinctively treat it like a traditional construction project. The mental model is straightforward: Design the blueprint (the functional specifications), build the walls (configure modules), install the wiring (develop integrations), and then, triumphantly, hand over the keys (go-live). In this paradigm, the project is deemed "done" and "successful" simply when the house is standing, regardless of whether it's functional, comfortable, or truly meets the family's needs.

But an SAP system is fundamentally not a house. It's far more akin to a living, evolving, highly interconnected ecosystem—a complex organism that must adapt, breathe, and grow with your business. You cannot adequately measure its success simply by ticking off whether each technical component is physically in place. Consider these common scenarios:

- A complex custom report is technically delivered on time, perfectly coded, but then sits unused and ignored by managers because it provides irrelevant data or is too difficult to interpret. This is not success.

- A new automated workflow is technically "live" in the system, but it's so clunky, counterintuitive, or riddled with exceptions that employees consistently bypass it, resorting to manual spreadsheets or unofficial processes. This is not success.

- A new integration point is technically functional, seamlessly transferring data between SAP and a legacy system, but the data itself is so dirty or inconsistent that the integrated information is unreliable and mistrusted. This is not success.

- The system goes live on budget and on time, but user adoption is low, and key business processes are slower or more complex than before. This is not success.

This prevalent checklist thinking leads to what I call the illusion of progress: a flurry of activity, significant expenditure, and countless completed tasks but, ultimately, very little tangible business impact or value realized. The project might meet its internal metrics for scope, schedule, and budget yet still be a strategic failure because it didn't move the needle on the core business challenges it was meant to address.

OUTCOMES AS THE TRUE NORTH: THE ULTIMATE MEASURE OF VALUE

So, if deliverables are merely means to an end, what then should be the true, ultimate measure of success for your SAP project? The answer is unequivocally outcomes. Outcomes are the tangible, measurable business results that truly matter to your organization, directly impacting profitability, efficiency, customer satisfaction, and strategic agility. They are the changes in business performance that justify the entire investment.

Outcomes are always tied to a specific business problem solved or a strategic opportunity realized. They are typically expressed in terms of

- Financial performance
  - Achieving a faster time to close the books at month-end (e.g., from ten days to three days)
  - Lowering inventory carrying costs by a measurable percentage (e.g., 15 percent) through optimized inventory management
  - Improving cash flow through accelerated collections or optimized payment terms
  - Increasing profitability by product line, customer segment, or geography by X percent due to better cost visibility
- Operational efficiency
  - Shortening the cycle time from quote to cash (e.g., from fifteen days to seven days) for sales orders
  - Reducing the number of manual work-arounds in core processes by 50 percent
  - Decreasing production downtime due to better visibility into material availability and machine maintenance
  - Reducing errors in invoicing, shipping, or payroll by a specific percentage
- Customer experience
  - Increasing on-time delivery rates to customers by a measurable amount
  - Improving customer satisfaction scores (CSAT or NPS) due to faster service, accurate orders, or better communication

- Reducing customer complaint resolution time by X hours/days
- Enabling new customer self-service capabilities that reduce calls to support centers

- Decision-making and insight
  - Providing real-time visibility into profitability by product, region, or customer, enabling faster, more informed strategic decisions
  - Eliminating reliance on multiple conflicting spreadsheets for critical financial or operational reporting
  - Enabling predictive analytics for sales forecasting, demand planning, or equipment maintenance

- Employee experience and productivity
  - Achieving significantly higher employee adoption and satisfaction with the new system
  - Reducing the time spent on manual data entry or reconciliation, freeing employees for higher-value tasks
  - Improving the speed and accuracy of internal cross-functional handoffs

- Strategic agility
  - Enabling faster integration of newly acquired businesses onto a standardized platform
  - Accelerating the launch of new products or services into the market
  - Supporting seamless global expansion into new countries or regions

Notice what is conspicuously missing from this list: "system configured on time," "integration delivered," "workflow approved," or "user training completed." Those are indeed important milestones and deliverables, but they are fundamentally not the ultimate destination. They are merely the means to achieve the desired business outcomes.

When leaders decisively shift their focus to outcomes, every other aspect of the SAP project—the specific deliverables, the project timelines, the allocated budgets, the choice of methodology—becomes a means to an end, a tool to achieve a strategic objective rather than an end in itself. This shift empowers the entire project team to prioritize activities that genuinely move the needle on business value.

THE CONSULTANT'S ROLE IN DELIVERABLES VS. OUTCOMES: A SHARED RESPONSIBILITY

Here's where clients, particularly those new to large-scale ERP implementations, often get tripped up. Consulting firms, by their very nature, training, and contractual obligations, are exquisitely good at delivering precisely what is explicitly stated in their contract. If your statement of work or project plan specifies a hundred custom workflows, they will deliver them. If you ask for ten specific reports, they will build them. Consultants, quite rightly, are incentivized and measured by whether they fulfill the defined deliverables within the agreed-upon scope, schedule, and budget.

But this inherent focus on deliverables creates a crucial blind spot unless the client actively intervenes. The question you, as the client leader, need to incessantly ask—of your consultants, of your internal team, and of yourself—is "Will these specific deliverables actually solve the fundamental business problem we set out to address, and will they contribute directly to our desired business outcomes?"

That's a question your consultants, by themselves, cannot fully answer. They can advise on technical feasibility, best practices, and alternative solutions. They can tell you how to build what you want. But they cannot define what problem truly needs solving or why a particular outcome is strategically paramount for your business. That profound insight and strategic direction require robust, engaged business leadership from the client side. It demands constant, proactive alignment between what is being delivered technically and why it is being delivered strategically. Without this client-driven outcome focus, consultants will efficiently build the house, but it might not be the right house for your family.

A TALE OF TWO PROJECTS: MEASURING ACTIVITY VS. IMPACT

Let me share two contrasting client stories, drawing from real-world experiences to illustrate the profound difference between a deliverable-driven approach and an outcome-focused one.

Client A: The Checklist Conquerors

Client A, a multinational consumer goods company, embarked on an SAP S/4HANA implementation driven by a project charter that heavily emphasized "completing all requirements" from various business units. Their approach was fundamentally that of a massive checklist. The project team—both internal and external consultants—defined thousands of granular requirements, meticulously documented each one, and relentlessly tracked their sign-off as a measure of progress. The consultants were judged almost purely on whether each requirement was technically delivered within the specified time frame.

At the project's conclusion, Client A proudly announced go-live. They had, by their metrics, successfully delivered an SAP system crammed with hundreds of custom workflows, thousands of bespoke reports, and dozens of unique tables and configurations. Every item on their original massive requirements checklist was "checked."

But the business was quickly overwhelmed. Employees, accustomed to their old ways, resisted the new, complex processes because they didn't clearly see how these new workflows improved their daily lives or aligned with any higher strategic goal. Reporting remained largely manual and inconsistent, despite the new system, because the data quality was poor and the multitude of custom reports created more confusion than clarity. Leadership, looking at the operational chaos and continued reliance on spreadsheets, saw little tangible value beyond a technical upgrade. The project was a costly technical success but a strategic business failure. They got what they asked for, but not what they truly needed.

Client B: The Outcome Achievers

Client B, a rapidly growing industrial manufacturer, took a radically different approach to their SAP implementation. From the very inception, their executive leadership defined just a few extremely clear and highly measurable core outcomes that would define success. These were:

- Shorten their order-to-cash cycle from an average of ten days to five days.

- Cut their month-end financial close time in half (from eight days to four days).

- Reduce duplicate data entry across all systems by 75 percent for key master data elements.

Throughout every phase of the project—from design workshops to testing cycles and go-live preparation—these three core outcomes served as their unwavering North Star. When consultants proposed new workflows or custom reports, Client B's leaders and business champions didn't just ask, "Can SAP do this?" They asked, directly and consistently, "Does this specific deliverable—this workflow, this report, this integration—directly help us achieve one of our three core outcomes?" If the answer was no, or if the link was tenuous, the proposed deliverable was rigorously challenged, simplified, or often, simply cut from scope. When trade-offs arose between competing demands or budget constraints, the decision was always made by referring back to which option best advanced the defined outcomes.

The result? Client B went live with far fewer deliverables than Client A—fewer custom reports, fewer complex workflows, and a much cleaner, more standardized system. But the impact was profound and immediate. They achieved all three of their targeted outcomes within six months of go-live, leading to significant cost savings, improved cash flow, greater customer satisfaction, and a highly engaged, empowered workforce that quickly adopted the new system. Their project was a true transformation, not just a technical upgrade.

HOW TO KEEP YOUR PROJECT OUTCOME-FOCUSED: PRACTICAL STRATEGIES

Shifting from a deliverable-centric to an outcome-focused approach requires intentional effort, discipline, and consistent leadership. Here are practical strategies to ensure your SAP project drives real value.

Define Outcomes Early and Clearly (and Keep Them Visible)

This is the foundational step. Don't wait until the testing phase, or worse, after go-live, to figure out what success looks like. At the absolute project kickoff, or even before, the executive steering committee must collaboratively define three to five (no more than seven) measurable, strategic business outcomes that will unequivocally define the SAP project's success. These outcomes should meet the following criteria:

- Be SMART: specific, measurable, achievable, relevant, and time-bound

- Be business-focused, not technical

- Be owned by business leaders, not just IT

Communicate these outcomes relentlessly. Post them in project rooms, include them in every presentation, discuss them in every meeting. They become the project's mantra and North Star.

Align Every Deliverable to a Specific Outcome

For every proposed workflow, every report design, every integration point, every piece of custom development, demand a clear answer to a simple question: "Which of our defined three to five strategic outcomes does this specific deliverable directly support?"

If the answer is none, or if the link is weak and indirect, then that deliverable is likely non-value-added and should be rigorously challenged, simplified, or cut from the scope. This creates a powerful strategic filter that purges unnecessary complexity and prevents scope creep.

Implement an "outcome mapping" exercise where every major functional requirement is directly traceable to one or more of your primary business outcomes. This forces clarity and accountability.

Use Outcomes to Drive All Decision-Making

Your defined outcomes are not just aspirational statements; they are your most powerful decision-making tools. When your project inevitably faces difficult choices—when budgets tighten, deadlines slip, or scope conflicts arise between departments—revisit your core outcomes.

- Prioritization: Which option or which deliverable best advances the outcomes you care most about?

- Trade-offs: If you can't achieve everything, what must be deferred or simplified to ensure you hit your most critical outcomes?

- Conflict resolution: Use the outcomes as an objective arbiter. Instead of relying on political power or loudest voice, the answer is derived from strategic alignment.

- Scope management: Any request to add scope must demonstrate a clear, direct, and significant contribution to a defined outcome. If it doesn't, the answer is no for the initial go-live.

Measure and Celebrate Outcomes, Not Just Deliverables

Shift your internal reporting and recognition mechanisms. Don't just hold a party because "all two hundred test cases passed" or "the system went live on time." While these are achievements, they are not the ultimate measure. Instead, redefine what success looks like by focusing your recognition on measurable business impact:

- Publicly measure and celebrate tangible business outcomes. Hold a celebration when "our order entry time just dropped by 40 percent," or "our month-end close is now three days faster," or "customer satisfaction scores are up 10 percent."

- Develop outcome-based dashboards. Create visual representations of your key outcomes (e.g., cycle times, cost reductions, adoption rates) that are accessible to all stakeholders.

- Connect individual contributions to outcomes. Help project team members and business users understand how their daily tasks directly contribute to achieving these larger business results. This builds morale and reinforces purpose.

Hold Consultants Accountable for Business Value, Not Just Tasks

Your relationship with your consulting partner needs to evolve from a transactional "fee-for-service" model to a true partnership where value realization is a shared objective. To build this kind of outcome-driven partnership, consider the following practical approaches:

- Incorporate outcome-based KPIs in contracts: If feasible, include specific business outcome KPIs in your consultant contracts or performance reviews. While not always easy to directly link, setting expectations around contributing to these outcomes changes the dynamic.

- Measure "move the needle": Beyond tracking whether deliverables were completed, regularly assess whether what they delivered actually "moved the needle" on your defined business outcomes. Engage in joint problem-solving if outcomes are not being met rather than simply accepting technical completion.

- Demand business context: Challenge consultants to explain how their proposed solutions or technical configurations will genuinely solve a business problem or enable an outcome. Don't let them retreat into purely technical language.

SHIFTING YOUR ORGANIZATION'S MINDSET: A CULTURAL TRANSFORMATION

This profound shift from a deliverable-centric to an outcome-focused approach is not merely a change in project management methodology; it requires a deep and sustained cultural transformation within your organization. Many companies are deeply ingrained in traditional project management thinking, where scope, timeline, and budget are the ultimate arbiters of success. While these are unquestionably important guardrails for project execution, they should serve as parameters for how you achieve success, not the ultimate definition of success.

To truly embed this outcome-focused mindset, you, as a leader, must set the tone, consistently and visibly. This begins with how you communicate. Weave outcome-focused language into every communication—emails, meetings, town halls, one-on-one discussions. Replace "What did we get done today?" with "What business problem did we solve today?"

When receiving project updates, press for clarity in terms of outcomes. Ask questions like "How is this contributing to reducing our month-end close time?" or "What impact is this having on customer satisfaction?"

Train your managers with the tools to reinforce mindset. Train them to ask, "So what?" Empower and train your managers to critically evaluate every activity and deliverable by asking, "So what? What's the business impact? What problem does this solve?" This cultivates a culture of accountability for value, not just activity.

Equally important is helping every team member see their role in the bigger picture. From the most junior user to the most senior executive, understand how their daily work, and the specific deliverables they contribute, link directly to the achievement of the larger business outcomes. This creates purpose and motivation.

Lastly, none of this matters if you don't walk the talk. As a leader, your actions speak louder than words. Demonstrate your commitment to outcomes by making decisions that prioritize them, even if it means saying no to a politically popular but non-outcome-driving deliverable.

This cultural shift takes time and persistence, but it is indispensable for unlocking the full transformative potential of your SAP investment.

THE ROLE OF METRICS: FROM ACTIVITY TO IMPACT

To effectively measure outcomes, you need robust and relevant metrics like those in the following list. This moves beyond simply tracking project activities (e.g., number of configurations completed, lines of code written) to measuring genuine business impact.

- Identify KPIs: For each of your three to five strategic outcomes, define specific, quantifiable KPIs that will allow you to track progress. For example, if "faster time to close the books" is an outcome, the KPI is "number of days to month-end close."

- Establish baselines: Before the project starts, meticulously measure the current performance of these KPIs. This "baseline" provides the starting point against which you will measure improvement.

- Set targets: Define clear, ambitious, but achievable targets for each KPI. For instance, "reduce month-end close from ten days (baseline) to four days (target)."

- Implement continuous monitoring: Set up dashboards and reporting mechanisms to continuously track these outcome-based KPIs, not just during the project but for months and years after go-live. This allows for real-time adjustments and demonstrates ongoing value realization.

- Differentiate leading vs. lagging indicators:

  - Leading indicators: Metrics that provide early warning signs or demonstrate progress toward an outcome before it's fully realized (e.g., user training completion rates, reduction in data errors during migration)

  - Lagging indicators: Metrics that show the ultimate business outcome after the change has taken effect (e.g., actual reduction in inventory costs, improved customer satisfaction scores). A healthy outcome-focused project uses both to manage progress.

OVERCOMING RESISTANCE TO OUTCOME FOCUS: ADDRESSING THE PUSHBACK

Implementing an outcome-focused approach often faces internal resistance, as it represents a significant shift from familiar ways of working. Common pushbacks include:

- "Outcomes are too hard to measure directly." Counter this by emphasizing that while direct causation can be complex, robust KPIs and clear baselines provide strong correlation and demonstrate impact. Start simple and refine.

- "That's not my job; that's the business's job." Remind teams that in a transformation, value realization is everyone's job. IT enables, but business champions must own the outcomes. This reinforces the shared responsibility.

- "We're too busy delivering features to worry about outcomes." This highlights the very problem: activity without purpose. Reiterate that the purpose of the features is the outcome. Connect their daily work to the bigger "why."

- "Outcomes can be manipulated." Establish clear governance and independent validation processes for outcome metrics to build trust and accountability.

Leading through this resistance requires persistence, clear communication, and a consistent demonstration of executive commitment to the outcome-driven approach.

SAP projects are simply too expensive, too disruptive, and too profoundly transformative to be measured merely by the weight of documents delivered, the number of workflows configured, or the adherence to a technical go-live date. These are critical steps, but they are not the finish line.

At the end of the day, no one truly cares whether your consultant delivered fifteen custom reports instead of twelve. What genuinely matters, what defines lasting success and value, is whether your finance team can now see real-time profitability clearly across all segments, whether your operational teams spend dramatically less time reconciling disparate spreadsheets, and whether your customers feel the tangible impact in faster, smoother, and more reliable interactions with your company.

When you strategically make outcomes your true north, relentlessly pursuing measurable business value, you fundamentally stop simply buying an implementation and start genuinely achieving transformation. And that—the profound and positive reshaping of your business for sustained competitive advantage—is the only fundamental reason to embark on a journey as ambitious and demanding as an SAP implementation in the first place.

Why Strategic Pushback Protects Your SAP Project

When an organization officially launches an SAP implementation project, the air is typically filled with optimism and, curiously, a tremendous amount of the word "yes."

You'll hear it from prospective partners in the sales cycle:

- "Yes, we can build that custom report to mirror your legacy system."

- "Yes, we can replicate that twenty-year-old intricate process exactly as you do it today."

- "Yes, we can configure the new system to precisely match the functionality of your old homegrown tool."

And you'll often hear it from your own internal teams, driven by comfort and a natural aversion to change:

- "Yes, we need that specific field or approval step for this one exception."

- "Yes, we must retain this unique reporting format because that's how we've always done it."

At first, this chorus of "yes" feels incredibly reassuring. After all, as a client, you naturally desire a partner who is accommodating, flexible, and capable of meeting your every request. But in the high-stakes, complex world of SAP projects, an endless, unchallenged series of "yes" answers is, counterintuitively, often the fastest and most insidious route to disaster. It's a path paved with good intentions but fraught with escalating costs, crippling complexity, and stifled innovation.

The most successful SAP clients—those who ultimately realize true business transformation—learn early and deeply that the word that most reliably saves projects, guards strategic intent, and ensures long-term agility is often a strategic, well-reasoned "no." This "no" isn't delivered because consultants are unwilling or inflexible; rather, it stems from their crucial role: to courageously challenge assumptions, to guard against shortsighted or suboptimal decisions, and to ensure the solution genuinely serves the future state of the business rather than merely perpetuating the outdated practices of the past. It is the protective friction that leads to clarity.

WHY CLIENTS STRUGGLE WITH "NO": THE PSYCHOLOGY OF RESISTANCE

It's an undeniable human tendency: Executives and business teams often interpret pushback—the word "no"—as resistance, a lack of capability, or even a personal affront. This perception makes it incredibly difficult for clients to accept, let alone embrace, dissent from their consultants. But in reality, a strategic no from an experienced SAP partner almost invariably comes from a place of deep experience, foresight, and a genuine desire to protect your long-term interests.

Consultants who have been through multiple SAP implementations have witnessed firsthand where certain paths inevitably lead. They have seen the painful consequences of

- Uncontrolled over-customization that cripples future upgrades, increases technical debt, and makes the system rigid and expensive to maintain

- Blindly copying broken or inefficient processes into a new, powerful system instead of seizing the opportunity to redesign and optimize them for the future

- Short-term decisions driven by immediate convenience or political pressure that ultimately cost millions in rework, missed opportunities, and operational friction down the line

- Fragmented decision-making that results in an incoherent system lacking a unified strategic purpose

A consultant who never says no to a client request is often one who is primarily protecting their contract, maximizing billable hours, or avoiding uncomfortable conversations rather than truly prioritizing your company's long-term health and strategic success. Such a consultant becomes an order taker, not a trusted advisor. Conversely, a consultant who demonstrates the courage and strategic insight to push back, to articulate the potential pitfalls, and to propose a better alternative is genuinely protecting your organization's future. They are choosing integrity over immediate gratification.

THE CONSULTANT'S PERSPECTIVE ON "NO": BEYOND COMPLIANCE

To truly understand the power of no, it helps to step into the consultant's shoes. Why would they risk client discomfort by saying it?

- Ethical responsibility: A reputable consulting firm views its role as being a strategic partner, not just a pair of hands. This implies an ethical responsibility to advise the client on the best path, even if that path is more challenging initially. Their long-term reputation rests on your long-term success, not just a quick go-live.

- Experience and foresight: Consultants leverage their experience across numerous projects. They have a "pattern recognition" capability—they've seen how specific types of customization or process replication have led to failure or crippling costs for other clients. Their "no" is often a preemptive warning based on hard-won lessons.

- Guarding SAP's integrity: A good consultant understands SAP's inherent design principles—its integration, standardization, and best-practice orientation. When a client requests something that fundamentally violates these principles and will render the system brittle, the consultant's pushback aims to preserve the integrity and long-term value of the SAP investment itself.

- Protecting future agility: They know that every nonstandard element makes future upgrades, maintenance, and the adoption of new SAP innovations more difficult and expensive. Their no is a vote for your future flexibility and lower total cost of ownership (TCO).

- Focus on value, not just features: A strategic consultant aims for your business outcomes, not just a checklist of features. If a request doesn't contribute to those outcomes, or worse, detracts from them, saying no is about guiding the project back to its core purpose.

WHEN "YES" IS DANGEROUS: THE PERILS OF CAPITULATION

There are pivotal moments in an SAP project when a "yes" answer from a consultant—or an internal project leader—is not a sign of collaboration or accommodation but rather an act of dangerous capitulation. These are the points where short-term appeasement trumps long-term strategic health, leading to crippling consequences:

- Saying yes to every customization request: When business users, resistant to adapting to new ways of working, insist on preserving every idiosyncratic legacy process or reporting format, and the project team or consultants simply acquiesce. This leads to an explosion of custom code (Z-programs), making the system unique, fragile, and incredibly expensive to maintain and upgrade. It digitizes inefficiency rather than transforming it.

- Saying yes to replicating obsolete reports: Leaders and users often demand that hundreds of legacy reports be replicated in SAP, without questioning their actual purpose, usage, or whether standard SAP reports or modern analytics tools could provide the same or better insights. Each unnecessary report adds development, testing, and maintenance overhead, bloating the system with "shelfware."

- Saying yes to manual work-arounds embedded in design: SAP is designed to automate and integrate processes. Saying yes to designing manual work-arounds into the new SAP system—rather than challenging and reengineering the underlying process—is a fundamental failure. It means you're spending millions on a powerful system only to force users to operate outside its integrated capabilities.

- Saying yes to undisciplined scope creep: When new requirements are continuously added without rigorous justification against the strategic vision or a proper change control process. Each yes to an additional feature, especially late in the project, introduces delays, budget overruns, and compounds complexity, diluting focus.

- Saying yes to unresolved conflicts: When departmental leaders have conflicting requirements and the project team (or consultants) fails to force a resolution, instead opting to build a solution that tries to accommodate all conflicting demands. This results in convoluted, inefficient processes that satisfy no one optimally and maintain internal friction.

Each seemingly innocuous yes in these scenarios feels like progress in the short term, as demands are met and tensions potentially diffused. However, each yes simultaneously layers on more complexity, escalates long-term costs, and introduces significant operational and technical risk. By the time of go-live, the system may look superficially familiar to the old one—but underneath, it's often fragile, bloated, out of sync with SAP's true strengths, and woefully unprepared for future demands.

CASE IN POINT: WHEN "YES" NEARLY SANK THE SHIP—A TALE OF TWO CLIENTS

Let's look at two contrasting scenarios that illustrate the dramatic difference a strategic "no" can make.

Scenario 1: The Manufacturer Overwhelmed by "Yes"

A midsize manufacturer, desperate to transition off an ancient, highly customized legacy system, began its SAP project with a mandate to minimize internal change. Business users, particularly in sales and customer service, insisted that the new SAP system precisely mimic every single field, every unique data element, and every specific screen layout from their twenty-year-old order entry system. Their chosen consultants, eager to avoid friction and ensure client satisfaction, largely said yes.

Months into the design and build phases, the consequences became dire. The project had accumulated over seven hundred custom fields across various modules. This led to

- Massive data migration issues: Reconciling legacy data into these bespoke fields was a nightmare.

- Inconsistent reporting: Data captured in custom fields often couldn't be easily integrated into standard SAP reports, leading to fragmented insights.

- A user experience worse than the old system: While visually familiar, the complex web of custom logic made navigation difficult and error prone.

- Exploding costs: Development, testing, and the inevitable rework for these customizations spiraled out of control.

Ultimately, the go-live was a partial success, but the system quickly proved unsustainable. The manufacturer later had to embark on another costly "simplification" project, realizing that the consultants' accommodating "yes" had created an unmanageable behemoth.

Scenario 2: The Client Empowered by "No"

Contrast that with a global chemicals distributor. They had a notoriously broken, multistep approval workflow for capital expenditures that often delayed critical investments by weeks. When discussing this with their SAP consultants, the internal team initially asked, "Can we just rebuild this exact workflow in SAP?"

The experienced consultants, understanding the principles of simplification and SAP's standard capabilities, courageously said no. Not a flat refusal but a reasoned pushback: "No, we can't rebuild that broken workflow. But here's a better way. We can redesign your approval process around SAP's standard workflow engine, which will reduce the number of steps, provide real-time visibility, and automate notifications."

Initially, there was discomfort within the client's finance and operations teams. It meant changing an established, albeit inefficient, habit. However, through diligent workshops led by the consultants, demonstrating the streamlined standard process and its benefits, the client agreed to adapt. The result? That initial no led to a 60 percent reduction in capital expenditure approval times. This seemingly small no translated into a massive win for efficiency, accelerating investment cycles and providing tangible, measurable business value.

The difference in outcomes was profound. In one case, accommodating "yes" led to complexity and cost. In the other, a strategic "no" paved the way for simplification and efficiency.

THE PSYCHOLOGY OF PUSHBACK: REFRAMING THE CONVERSATION

As a client leader, it's natural to feel a degree of discomfort or even irritation when your consultants tell you no. However, cultivating a successful SAP partnership requires you to reframe your perception of pushback:

- "No" is rarely a rejection of your business needs. Instead, it's usually a strategic redirect: away from shortsighted demands or outdated practices and toward solutions that genuinely serve the project's outcomes and your long-term strategic goals.

- It signals a consultant is thinking strategically, not just tactically. A consultant who only says yes is likely focused on completing tasks. One who says no is considering the broader impact on your business agility, TCO, and future innovation.

- It's a form of protection. When you hear no, instead of feeling frustrated, ask yourself: *What risk are they protecting us from? What long-term pain are they helping us avoid? What better alternative do they see that we might not?*

Embracing this perspective transforms "no" from a barrier into a valuable signal, an indicator of a truly invested and capable partner.

WHEN "YES" IS DANGEROUS: RED FLAGS IN DISGUISE

There are situations in an SAP project where a consultant's "yes" is not a sign of accommodation but rather a warning siren of impending trouble.

These are moments when leadership must be acutely aware and ready to intervene:

- To every user request for customization: When consultants readily agree to every business user's demand for bespoke functionality, especially when it stems from a desire to maintain comfort rather than create strategic value. This signals a lack of discipline and an openness to technical debt.

- To replicating legacy reports: When a proposal includes rebuilding hundreds of old, rarely used, or redundant reports in SAP without a rigorous challenge to their necessity or purpose. This is a waste of resources and bloats the system.

- To manual work-arounds within the new system: If a design proposes embedding manual, nonintegrated steps into a process that SAP is designed to automate, it indicates a failure to reengineer processes and leverage the system's core capabilities.

- To uncontrolled scope creep: When consultants accept every new requirement or change request without rigorous justification, impact analysis, or a formal change control process. This can quickly derail timelines and budgets.

- To compromise over clarity: When faced with conflicting stakeholder demands, consultants opt for a complex, middle-ground solution that tries to please everyone but sacrifices simplicity and strategic alignment.

- To ignoring data readiness: If consultants proceed with design and build without pushing back forcefully on poor master data quality or the lack of a robust data governance plan. They are effectively building on quicksand.

Each of these "yeses" feels like progress in the immediate term—a requirement fulfilled, a conflict avoided. But each layers on complexity, accelerates costs, and multiplies risks. By go-live, the system may be familiar, but it will be fragile, bloated, and out of sync with SAP's strengths, delivering only a fraction of its potential value.

BUILDING TRUST TO EMBRACE "NO": THE FOUNDATION OF TRUE PARTNERSHIP

For pushback—the strategic "no"—to function effectively, a foundation of deep trust between the client organization and its consulting partner is absolutely nonnegotiable. Clients need to genuinely believe that consultants are not being obstinate or inflexible for ego's sake but are instead steering them toward long-term value and protecting them from future pain. The most productive SAP relationships are characterized by

- Transparency and explanation from consultants: When a consultant says no, they must always follow it with a clear, concise, and business-focused explanation of why. They should articulate the risks of the proposed path (e.g., "This customization will make your future S/4HANA upgrade impossible") and present viable, strategic alternatives (e.g., "Instead, let's reengineer this process to leverage standard SAP best practices, which will provide X, Y, Z benefits").

- Openness and active listening from clients: Clients must cultivate an environment where they are open to alternatives and willing to challenge their own ingrained assumptions and legacy habits. This means actively listening to the consultant's rationale, engaging in constructive debate, and being prepared to adapt internal processes for the greater good of the transformation.

- Shared anchor in business outcomes: Both sides—client and consultant—must consistently anchor all design decisions, all debates, and all instances of "yes" or "no" to the clearly defined strategic business outcomes (as discussed in Chapter 11). This shared purpose provides an objective framework for resolving conflict and ensures that choices are made based on what truly moves the business forward, not comfort zones or historical inertia.

- Mutual respect and humility: Both parties must approach the relationship with respect for each other's expertise. Consultants respect the client's deep business knowledge; clients respect the consultant's SAP expertise and cross-industry experience. Humility allows both sides to admit when their initial assumptions might be flawed and to adapt for the project's success.

Without this mutual trust and shared understanding, a consultant's "no" will simply sound like resistance, generating friction and undermining the very partnership meant to drive the transformation. With trust, "no" sounds like protection, a critical guidepost on a complex journey.

PRACTICAL WAYS CLIENTS CAN LEVERAGE "NO": EMPOWERING STRATEGIC PUSHBACK

As a client leader, you have the power to fundamentally change the dynamic and proactively invite and leverage the invaluable insights that strategic pushback provides. Here's how:

- Explicitly invite pushback early: From the very first kickoff meeting, tell your consultants, clearly and unambiguously, "We expect you to be our strategic advisors. If you believe we are proposing something that is harmful to our long-term interests, or if you see a better way, we expect you to challenge us. Your courage to say no is a sign of true partnership." This sets the expectation and creates psychological safety for consultants.

- Always ask for alternatives (and their implications): When you hear no to a request, immediately follow up with "Okay, I understand. What is a better way to achieve this business outcome using standard SAP or a simpler approach? What are the trade-offs of that alternative? What are the long-term benefits and costs?" A good "no" should always be followed by a valuable "here's a better way."

- Don't punish honesty; reward it: If a consultant saves you from a costly mistake by saying no, even if it initially causes internal discomfort or a slight delay, acknowledge and reward that honesty. Publicly praise consultants (and your internal team members) who demonstrate courage in challenging the status quo for the project's greater good. This reinforces the desired behavior and fosters a culture of transparency.

- Look for patterns in pushback: If you consistently hear no around the same themes (e.g., excessive customization, complex reporting requirements, resistance to process redesign), it's a powerful signal. This isn't just about individual requests; it indicates deeper strategic misalignments, unresolved internal conflicts, or a fundamental misunderstanding of the SAP paradigm within your organization. Use these patterns to trigger strategic interventions at the executive level.

- Utilize governance boards wisely: Your project steering committee or design authority board should be the forum where these "no" decisions are formally reviewed, openly debated, and strategically ratified. It should not be a place where a strategic "no" from the project team is simply overridden by political pressure or personal preference. The board's role is to uphold the strategic vision and make decisions that align with long-term outcomes, even if difficult.

THE COST OF IGNORING "NO": THE ECHO OF REGRET

When clients disregard the strategic "no" and choose to bulldoze past the warnings of their experienced consultants, the consequences are stark, predictable, and deeply painful:

- Massive budget overruns: Every unnecessary "yes" to customization or complexity directly inflates development, testing, and support costs beyond original estimates.

- Crippling project delays: Complexity multiplies testing cycles, debugging efforts, and the overall time required to get the system operational and stable.

- Broken upgrades and stifled innovation: Heavily customized systems become brittle. Major SAP innovations and new functionalities often cannot be adopted easily or at all, as your system is too tangled, leaving you increasingly reliant on outdated, expensive bespoke solutions.

- Loss of agility and responsiveness: A complex, customized system is inherently rigid, unable to adapt swiftly to changing market conditions, new business models, or competitive threats.

- Demoralized workforce and low adoption: Users grow frustrated with clunky, inefficient processes or systems that don't support their work, leading to active resistance and a failure to realize the expected benefits.

- Technical debt mounts: What looks like short-term "progress" with a series of "yes" answers quickly accumulates into massive technical debt that your business will continue to pay for, year after year, long after the initial implementation is completed.

REFRAMING PUSHBACK AS A VALUE SIGNAL: THE SENTINEL OF SUCCESS

Ultimately, shift your perspective. Think of a consultant's strategic "no" not as a roadblock to your immediate desires but as a clear, loud signal of value alignment. When viewed through this lens, a well-founded no from your consulting partner can signify several important things:

- It demonstrates that your consultants are deeply engaged, thinking critically, and acting as true partners, not just order takers.

- It means they are treating your business's long-term health and strategic success as if it were their own.

- It serves as a powerful reminder that the fundamental point of implementing SAP is not to preserve comfort or replicate the past but to achieve genuine, profound transformation and unlock future potential.

Embracing strategic pushback is a mark of a mature, outcome-driven organization. It differentiates those who merely implement software from those who achieve lasting business impact.

The most successful SAP leaders I have encountered share a common, albeit initially counterintuitive, trait: They develop a healthy, appreciative relationship with the word "no." They understand that it is the necessary friction that sharpens their strategic intent, the vital guardrail that prevents disastrous detours, and the unvarnished honesty that keeps a complex project grounded in reality.

When your trusted SAP consultants say no to a request, listen carefully and engage deeply. Because in the complex landscape of an SAP transformation, the "nos" you courageously accept and act upon today often pave the way for the resounding "yeses" you will celebrate tomorrow, like these:

- Yes, the system runs smoothly and efficiently, year after year.

- Yes, users quickly adapt and confidently leverage the new, streamlined processes.

- Yes, future SAP upgrades are simple, predictable, and cost effective.

- Yes, you achieved the profound business transformation and strategic agility you set out to realize.

In the end, the power of "no" is not about restriction; it is about liberation. It is the quiet, often unglamorous, yet absolutely essential engine of lasting SAP success, ensuring your transformation delivers clarity, not complexity.

Speed Without Direction: The Hidden Cost in SAP Projects

In today's relentless business world, speed is often lauded, almost fetishized, as the ultimate competitive advantage. Leaders are under immense pressure to launch products faster, streamline processes quicker, and fundamentally transform their organizations at an ever-accelerating pace. When it comes to a monumental undertaking like an SAP implementation, this same pervasive pressure applies with intense force: Boards demand rapid returns on investment, executives clamor for a swift go-live, and project teams are relentlessly pushed to "move fast" in the name of efficiency and agility. The idea of a "fast-track" implementation holds immense appeal.

But here lies a critical and often overlooked truth: Speed without clear, strategic direction is akin to driving a high-performance Formula 1 car without a GPS, a map, or even a determined destination. You'll undoubtedly go somewhere incredibly quickly, generate a lot of impressive motion, and burn through significant fuel (resources), but there's no guarantee—indeed, a high probability—that you'll end up anywhere near where you actually need to be. Many SAP projects, despite their ambitious beginnings and rapid initial pace, ultimately falter not because they moved too slowly but because they moved with breathtaking velocity in the wrong direction. The ultimate purpose of this chapter is to equip you, as a client leader, with the foresight and conviction to recognize that clarity of vision and strategic direction matter infinitely more than mere velocity in an SAP project and to show you how to ensure your transformation maintains the right trajectory from its beginning to its successful conclusion.

THE ILLUSION OF SPEED: DECONSTRUCTING THE MYTH OF "FAST-TRACK" SUCCESS

Executives often boast about achieving "fast-track SAP implementations," pointing to shortened timelines and rapid go-lives as irrefutable proof of project success. On the surface, such speed sounds undeniably appealing. In some specific, highly controlled scenarios (e.g., very small, greenfield implementations with minimal complexity and high standardization), speed can indeed deliver legitimate benefits: reduced overall project costs, less fatigue from prolonged, multiyear projects, and earlier visibility into new system capabilities.

However, in the vast majority of complex enterprise SAP transformations, this celebrated speed is often nothing more than a dangerous illusion. What appears to be a quick win frequently masks deep-seated problems that will inevitably surface later, often with compounding costs and significant pain. There are a number of realities to consider, the first being a sense of false velocity and real debt. A project that proudly declares itself "live in nine months" instead of a more realistic twelve or eighteen, but then requires two years of intensive, expensive post-go-live remediation, endless bug fixes, and re-implementing poorly designed processes, is not fast. It's simply accumulated technical debt and operational debt at an accelerated rate, leaving your organization crippled.

The second reality is a project team that rushes to configure SAP modules and develop custom code without dedicating adequate time and resources to crucial activities like comprehensive change management, thorough user training, and rigorous end-to-end business process testing is not efficient; it's simply generating activity without meaningful impact. The system may be technically functional, but it will be functionally ineffective, leading to low user adoption and continued reliance on shadow systems.

The third realist is the facade of "go-live." A project that "goes live" on time and on budget but leaves frontline users confused, frustrated, and unable to perform their daily tasks efficiently is not a success; it's a costly facade. The project might hit a technical milestone, but it fails to deliver the promised business value. You've moved quickly to a destination that no one wants to inhabit.

Lastly, the hidden costs explode later. The shortcuts taken in the name of speed during the initial phases—skimping on data cleansing, deferring critical integration testing, rushing through blueprinting—do not eliminate work; they merely defer it. This deferred work inevitably surfaces as massive problems after go-live, requiring emergency fixes, additional budget allocations, and frantic problem-solving that are exponentially more expensive and disruptive than addressing issues proactively up front.

In essence, speed that fundamentally sacrifices clear direction, meticulous alignment, and genuine organizational readiness does not lead to less waste; it leads to significantly more waste, more frustration, and ultimately, a much longer, more painful journey to value realization. The allure of speed can blind leaders to the critical foundational work that guarantees sustainable success.

WHY DIRECTION MATTERS MORE: THE STRATEGIC IMPERATIVE

If speed without direction is perilous, then direction is the indispensable strategic North Star of your entire SAP project. It is the compass that guides every decision, prioritizes every activity, and ensures that the immense investment of time, money, and human capital is channeled toward precisely where your business needs to be. Direction answers these most fundamental questions:

- What specific, measurable business outcomes are we truly trying to achieve with this SAP transformation? Is it a 15 percent reduction in inventory carrying costs, a 50 percent faster financial close, or a 20 percent improvement in customer on-time delivery?

- How will SAP help us authentically deliver on our core customer promises and enhance their experience? Will it enable faster service, more accurate orders, or personalized interactions?

- What existing business processes must be rigorously harmonized, simplified, or completely reengineered before they are automated in SAP? Are we simply digitizing chaos or creating efficiency?

- How do we ensure this transformation strategically positions us for the next decade of growth and innovation, not just for the next fiscal quarter's reporting requirements? Is it built for agility, scalability, and adaptability?

- What is our unique competitive advantage, and how will SAP amplify it rather than dilute it by forcing us into generic "best practices" that don't fit our niche?

Without clear, unwavering direction, speed simply amplifies mistakes. An ancient proverb wisely states, "If you don't know where you're going, any road will take you there." In the context of an SAP implementation, this means you will consume vast resources traveling aimlessly, accumulating features without purpose, and building a system that may be technically functional but strategically irrelevant. SAP is far too significant an investment—financially, organizationally, and strategically—to allow the powerful, yet often deceptive, siren song of speed to override the paramount importance of a clear, disciplined vision.

COMMON PITFALLS WHEN SPEED TAKES OVER: THE COST OF SHORTCUTS

When speed becomes the primary, unchecked driver of an SAP implementation, it inevitably leads to a series of common, yet incredibly costly, pitfalls. These are the "shortcuts" that almost always result in longer, more expensive, and less impactful outcomes:

- Shortcutting design and process reengineering

  - The trap: Rushing through critical design workshops (blueprint, fit-to-standard) means superficial analysis of existing processes and a failure to challenge legacy inefficiencies. The easy path is taken: simply re-implementing your exact twenty-year-old processes in SAP, rather than optimizing them.

  - The cost: This creates a shiny new system that merely replicates yesterday's inefficiencies. You lose the opportunity for true process harmonization, standardization, and optimization that SAP offers. This often leads to excessive, unnecessary customization and a system that is rigid and difficult to adapt to future needs.

- Skipping or minimizing change management

  - The trap: Activities related to change management, communication, stakeholder engagement, and user adoption are frequently viewed as "soft," "nontechnical," and therefore easily pushed aside or drastically minimized in the name of speed. "We'll do training right before go-live, that's enough."

- The cost: The result is predictable: a chaotic go-live, pervasive user resistance, low adoption rates, and the proliferation of "shadow systems" (manual spreadsheets, offline databases) as employees try to work around the new system. The business fails to realize value because people don't use the system as intended, leading to continued inefficiencies and frustrated employees.

- Ignoring or deferring data quality and governance

  - The trap: Cleansing and harmonizing master data (customers, products, vendors, finance data) is tedious, time-consuming, and often seen as a pre-project chore. Organizations underestimate the effort and push it into or past the build phase. "We'll fix the data later."

  - The cost: Moving "bad data fast" simply creates bad outcomes faster (as detailed in Chapter 5). Inaccurate financial reporting, incorrect inventory, delayed orders, and frustrated customers are direct consequences. Data problems are the single biggest cause of post-go-live operational chaos and a fundamental undermining of user trust in the new system.

- Over-customizing for quick wins (short-term comfort)

  - The trap: To appease impatient business demands and avoid internal friction during a fast-paced build, consultants (often implicitly encouraged by the client's lack of firmness) may extensively customize the system. It feels faster in the moment because it avoids difficult conversations about process change.

  - The cost: As discussed in Chapter 7, this creates crippling long-term complexity and technical debt. Upgrades become nightmares, maintenance costs skyrocket, and the system's ability to adopt future innovations is severely restricted, leaving the organization technologically and strategically stagnant.

- Losing sight of end-to-end integration

  - The trap: In a desperate rush to configure individual SAP modules or satisfy departmental requirements independently, organizations miss the overarching importance of end-to-end process integration across functions. The focus is on making specific pieces work rather than the entire business flow.

  - The cost: The project may technically "work in silos"—Sales can create orders, Finance can post invoices—but it fails as a true integrated business solution. Data doesn't flow seamlessly, manual reconciliation points remain, and the core promise of an integrated ERP is fundamentally undermined. This defeats the primary purpose of SAP.

- Inadequate Testing and Quality Assurance:

  - The trap: Under pressure for speed, testing cycles are often truncated, less rigorous, or conducted with insufficient business user involvement. Critical end-to-end scenarios are skipped, and performance testing is often neglected.

  - The cost: This leads to a go-live fraught with unexpected bugs, system performance issues, and broken business processes. The project team transitions from build mode to perpetual fire-fighting, impacting business continuity and user confidence.

- Weak governance and decision-making

  - The trap: When speed is paramount, governance committees might rubber-stamp decisions or defer critical trade-offs, allowing quick, tactical choices to override strategic oversight.

  - The cost: This allows conflicting priorities and unaligned decisions to permeate the system design, leading to a fragmented, inefficient SAP solution that fails to serve the overarching business strategy.

BALANCING SPEED AND DIRECTION: PRINCIPLES FOR NAVIGATING THE TENSION

To be absolutely clear: Speed is not the enemy. Projects that endlessly meander, stretching for years with no end in sight, can indeed kill momentum, exhaust resources, and erode stakeholder confidence. The true challenge for leaders is to find the crucial balance between disciplined execution velocity and unwavering strategic direction. There are a few core principles that will ensure direction remains the absolute priority.

First, anchor everything to measurable business outcomes. This is your unbreakable compass. Every single decision made during the SAP project—every configuration choice, every process design, every customization request, every resource allocation—must be rigorously evaluated against the strategic business outcomes of the project. Ask constantly, "Will this choice move us demonstrably closer to our goal of a simpler process, a better customer experience, a faster financial close, or improved regulatory compliance?" If the answer is no, or if the link is tenuous, then that activity or deliverable is likely noise and should be challenged or cut.

Second, embrace the strategically aligned MVP mindset. Instead of racing toward an all-or-nothing, "big bang" go-live that attempts to do everything at once, think in terms of incremental, value-driven releases (minimum viable products, or MVPs). The key is that each increment must clearly align with, and build toward, the long-term strategic vision—it shouldn't just deliver features for the sake of immediate speed or simply moving fast. A strategically aligned MVP delivers early, tangible business value, builds momentum, and allows for iterative learning and refinement without compromising the overall direction.

Equally important is governance with unwavering discipline. Establish a powerful, engaged steering committee or governance board that consistently tests speed-driven decisions against the project's foundational vision and defined outcomes. This committee's role is to act as the ultimate guardian of the strategic direction, empowered to say no to shortcuts that compromise long-term value, even when faced with immense pressure for velocity. They are the arbiters of strategic alignment.

At the same time, prioritize and integrate change management from day one. Don't view change management activities (communication, training, user engagement) as "soft" extras that can be deferred. Build dedicated time and resources for proactive adoption strategies directly into your project schedule, from the very beginning. Speed without widespread user adoption is merely a facade; the project might look live, but the business will still be operating in fragmented, old ways, undermining the entire investment. Proactive engagement mitigates resistance and accelerates genuine value realization.

Another essential principle is investing in data early and continuously. Recognize that robust data is the fundamental bedrock of any successful SAP system. Initiate master data assessment, cleansing, and governance efforts before you accelerate into configuration and development. Bad data will undermine both your speed (through constant rework and debugging) and your strategic direction (through unreliable reporting and flawed decision-making). A clean data foundation enables true speed and accurate navigation.

Finally, build in strategic pause points and review gates. Schedule deliberate moments within the project timeline for reflection, reassessment, and realignment. These "pause points" allow leadership to objectively evaluate progress against outcomes, identify emerging risks, and adjust direction, if necessary, before issues compound. This disciplined approach prevents blind acceleration down the wrong path.

THE INDISPENSABLE ROLE OF LEADERSHIP: GUARDIAN OF THE VISION

As the client leader, your role in an SAP transformation is paramount: You are the ultimate guardian of the vision and strategic direction. Consultants, project managers, and even your own internal teams may, consciously or unconsciously, push for shortcuts and compromises in the name of speed, immediate task completion, or perceived efficiency. But you must possess the courage and conviction to use questions like those below to continually bring the conversation back to the fundamental direction:

- "Are we aligning every decision with our overarching business strategy and the outcomes we defined?"

- "Are we genuinely building a resilient foundation for the next ten years of our business, not just racing to meet a deadline in the next ten months?"

- "Are we willing to strategically slow down where it absolutely matters—in design, data, or change management—to avoid exponentially more expensive, time-consuming rework and potential project failure later?"

- "Are we measuring actual business outcomes, or just activity and deliverables?"

Leadership in an SAP project means knowing precisely when to accelerate and, more critically, when to apply the brakes. It requires the wisdom to understand that sometimes, the bravest, most strategic, and ultimately fastest decision you can make is to say, "Not yet. Let's pause, ensure alignment, and get this right first." This disciplined approach prevents rushing to failure.

CASE IN POINT: THE COST OF LOSING DIRECTION—REAL-WORLD CONSEQUENCES

Let's examine two contrasting examples that starkly illustrate the long-term consequences of prioritizing speed over direction.

Scenario 1: The Global Manufacturer That Rushed to Failure

A large global manufacturing company, facing intense pressure to hit a specific corporate IT consolidation deadline, rushed their SAP S/4HANA implementation project with an aggressive "go-live in under a year" mandate. In their frantic pursuit of speed, the project team made critical compromises: They skipped comprehensive master data cleansing, minimized change management activities, and significantly deferred end-to-end integration testing, opting for isolated unit testing. The executive sponsor frequently emphasized "just get it live," allowing the project to prioritize technical completion over strategic alignment.

They technically met their deadline—a public announcement was made, and celebrations ensued. However, the subsequent go-live was catastrophic and profoundly disruptive.

- Customers experienced massive order delays and mispricing, leading to significant revenue loss and customer churn.

- Employees were overwhelmed and resorted to widespread use of manual spreadsheets and unofficial shadow systems to work around the dysfunctional SAP processes, completely undermining the system's purpose.

- Financial reporting was riddled with inaccuracies, leading to intense manual reconciliation and a lack of trust in the numbers.

- Operational efficiency plummeted, impacting production schedules and inventory accuracy.

What was initially celebrated as a "fast win" quickly devolved into a multiyear, multimillion-dollar remediation effort—costing far more than the original implementation—and a significant blow to executive credibility. Their speed had accelerated them straight into a crisis.

Scenario 2: The Thoughtful Client That Realized Value Sooner

Contrast this with a prominent pharmaceutical distribution client. They also faced pressure but chose a different path. Their leadership prioritized direction and quality over a rigid, arbitrary deadline. They deliberately took an additional six months in the initial phases to meticulously standardize core business processes across their regional entities, invest heavily in comprehensive master data cleansing and establishing robust data governance, conduct extensive, hands-on user training and involve key business users in iterative design and testing, and build a strong, two-way communication plan that prepared the organization for change.

Their go-live was not the "fastest" in their industry, but it was remarkably smooth, predictable, and highly successful. User adoption was exceptionally high, and employees quickly became proficient in the new system. Most importantly, the company achieved all of its key measurable return on investment (ROI) outcomes—including significant reductions in order fulfillment errors, a faster financial close, and improved supply chain visibility—within the first year of go-live. They were not the fastest to technically go live, but they were, unequivocally, the fastest to realize profound business value and sustained competitive advantage. Their strategic patience paid dividends.

The lessons from these experiences are clear and imperative for any leader embarking on an SAP journey:

- Speed is seductive but dangerous. True SAP success depends far more on clear, unwavering strategic direction than it does on mere velocity.

- Shortcutting leads to rework. Rushing through foundational phases (design, data, change management) does not save time; it merely creates immense technical debt, user adoption challenges, and ultimately, significantly more wasted effort and expense.

- Business outcomes are your compass. Every decision, every resource allocation, and every phase of the project should be directly and explicitly aligned with your predefined, measurable, long-term strategic goals and business outcomes.

- Leaders must fiercely protect the vision. Your fundamental role is to act as the ultimate guardian of the strategic direction, consistently keeping consultants and internal teams aligned, even when faced with intense pressure to "just go faster."

In a business world often obsessed with immediate gratification and breakneck speed, real leadership in an SAP transformation is about cultivating clarity, exercising strategic conviction, and demonstrating profound patience. SAP projects are not short-distance sprints where the first one across the finish line automatically wins. They are complex, enduring marathons that demand careful pacing, immense endurance, unwavering discipline, and, most critically, an absolute, unshakable sense of direction.

The next time someone in your organization asks, perhaps with urgency, "How fast can we go live?", your considered and strategic response should be, "As fast as we possibly can while staying absolutely true to our vision, our strategic outcomes, and the long-term health of our business." Because in the end, it's not the project that technically finishes first that truly wins; it's the project that authentically delivers the profound, lasting business outcomes your organization actually needs to thrive and innovate for decades to come.

Documentation Is a Superpower

If there's one word in the lexicon of business projects that consistently makes people roll their eyes, emit a collective sigh, or groan inwardly, it's arguably *documentation*. For many executives, it conjures images of tedious administrative overhead, a bureaucratic necessity that drains resources without delivering tangible value. Project teams frequently view it as an unwelcome burden, something that slows down their agile pace and diverts precious time from "real" work like configuration and testing. And consultants, often under pressure to hit aggressive deadlines, might be tempted to skip it entirely, rationalizing that "we'll remember how this works later" or "the client can always call us if they have questions."

But in the complex, high-stakes world of SAP implementation, this perception is a dangerous delusion. Documentation is not an afterthought; it is a profound superpower. When meticulously planned, rigorously executed, and continuously maintained, effective documentation becomes the invisible, yet indispensable, glue that holds a successful SAP transformation together. It acts as your organization's living institutional memory, protecting your massive investment, profoundly empowering your people, and preventing countless costly missteps long after the external consultants have packed up and moved on. It is the silent, unsung hero of enduring SAP success.

WHY DOCUMENTATION GETS OVERLOOKED: THE TEMPTATION OF SHORT-TERM GAINS

Given documentation's critical importance, why is it so frequently neglected, marginalized, or simply pushed to the very bottom of the priority list during SAP projects? Several powerful, yet misguided, reasons contribute to this common oversight.

The first of these is the relentless rush to go live, also known as "the finish line fallacy." SAP projects are often defined by intense pressure to hit specific go-live dates. The project team's collective energy and focus become overwhelmingly consumed by urgent, visible activities like configuration, development, and testing. Documentation, by its nature, feels less immediate, less "hands-on," and doesn't directly contribute to the technical act of flipping the switch. Consequently, it routinely slips to the bottom of the priority list, assumed to be something that can be caught up on "later." This "later" rarely, if ever, comes, as new priorities emerge post-go-live.

The second is consultant dependency and the "knowledge is portable" myth. Clients often assume that because external consultants are the ones designing and configuring the system, they will always be available to explain how things work, clarify complex configurations, or troubleshoot issues. This creates a dangerous dependency. Consultants, being external resources, eventually move on to other projects or even different firms. When key consultants or project team members leave without robust documentation in place, a substantial portion of institutional knowledge—such as the rationale behind key decisions, the design logic of the system, and the intent of specific configurations—leaves with them. This leaves the client vulnerable and reliant on costly reengagement.

Third is perceived complexity and overwhelm. SAP systems are vast, intricate, and deeply interconnected. The sheer scope of documenting every single detail can feel overwhelmingly daunting to a project team already stretched thin. The magnitude of the task often leads to procrastination or outright avoidance, with individuals telling themselves, "It's too big; we'll get to it when we have more time." The reality is, time rarely materializes for this catch-up.

Next is cultural undervaluing through documentation as bureaucracy. In many organizations, there's a deeply ingrained cultural perception that documentation is merely bureaucratic overhead—a necessary evil for compliance but not a value-add activity. Leaders may explicitly or implicitly communicate this by not allocating sufficient resources, not holding teams accountable for documentation quality, or not recognizing its strategic importance. This cultural undervaluing propagates down to project teams, further deprioritizing its creation.

Additionally, there is often too much focus on the "what" instead of "why." Project teams often document what was configured or developed but fail to capture the critical "why" behind those decisions. Without the strategic context of why a particular process was designed, why a specific configuration was chosen over another, or why a customization was justified, documentation becomes a mere technical reference that lacks meaning and makes future changes risky.

Last is the "tribal knowledge" syndrome. Over time, knowledge about how the system really works, its quirks, and its work-arounds resides solely within the heads of a few key internal "super users" or IT experts. This "tribal knowledge" is highly fragile. If those individuals leave, the organization faces a catastrophic loss of institutional memory, bringing processes to a halt.

DOCUMENTATION AS A STRATEGIC ASSET: YOUR INSURANCE POLICY AND OPERATING MANUAL

When viewed through a strategic lens, documentation transforms from a perceived burden into a powerful organizational superpower—your essential insurance policy, your comprehensive operating manual, and your invaluable enabler for continuous evolution. It's the guarantee that your massive SAP investment isn't just working today but can adapt, grow, evolve, and sustain your business operations effectively for years, if not decades, into the future.

Here's why truly comprehensive and well-maintained documentation is a strategic asset:

- Protects your investment and prevents knowledge loss: An SAP project typically represents millions, sometimes hundreds of millions, of dollars in investment. Documentation ensures that the critical knowledge, design rationale, and operational understanding underpinning that massive spend doesn't evaporate when key project team members, internal experts, or external consultants inevitably leave the organization. It safeguards your intellectual capital and maintains continuity.

- Empowers business users and accelerates adoption: Clear, concise, and accessible documentation (e.g., user guides, process flows) gives employees the confidence and autonomy to use the new SAP system as intended. It reduces frustration, minimizes the need for constant hand-holding, and prevents users from creating inefficient, nonstandard work-arounds because they don't understand the "right" way to operate. It empowers self-service and builds trust in the system.

- Enables continuous improvement and future agility: Without accurate documentation, every future change, every new requirement, every strategic enhancement becomes a costly and risky "rediscovery" mission. Documentation provides the baseline. It allows future teams to understand the existing landscape, build intelligently on what already works, and make informed decisions for process optimization or system enhancements. It's the foundational map for scaling operations, integrating new businesses, and adopting future SAP innovations.

- Reduces risk and ensures compliance: Clear, auditable records of your SAP configurations, security settings, process flows, and master data definitions are absolutely critical for regulatory compliance (e.g., SOX, GDPR, industry-specific regulations). Well-documented systems help your business withstand internal and external audits, mitigate compliance gaps, and avoid significant fines or reputational damage. It provides traceability and accountability for critical business processes and data.

- Facilitates faster troubleshooting and issue resolution: When a problem arises in SAP (e.g., an error in a transaction, a report generating incorrect data), comprehensive documentation helps IT support teams and business users quickly pinpoint the root cause. Instead of a lengthy investigation trying to decipher custom code or understand undocumented configurations, relevant documentation speeds up resolution, minimizing costly downtime and operational disruption.

- Reduces consultant dependency and cost: By maintaining robust internal documentation, you significantly reduce your ongoing reliance on expensive external consultants for routine support or minor changes. Your internal teams become more self-sufficient, leading to substantial long-term cost savings.

WHAT GOOD DOCUMENTATION LOOKS LIKE: MORE THAN JUST PDFS

Not all documentation is created equal. Simply dumping configuration screenshots into a disorganized PDF or a chaotic shared drive is woefully insufficient. Good documentation is a living, strategic asset that is thoughtfully designed and continuously maintained.

Good documentation should embody these core characteristics:

- Accessible: It must be stored in a centralized, easily searchable repository (e.g., a knowledge base, wiki, or document management system) that is readily available to all relevant stakeholders—not buried on someone's hard drive or in an obscure network folder. It should be easy for anyone who needs it to find what they're looking for, quickly.

- Understandable: It must be written in clear, plain language, avoiding excessive technical jargon wherever possible. It should provide not just what was configured or how a process works but crucial context on the "why"—why certain decisions were made, what business problem it solves, and what alternatives were considered. It should be tailored to the audience (e.g., business-friendly process maps for users, more technical details for IT).

- Up-to-date (living documents): Documentation is not a static, one-time deliverable at go-live; it is a living set of documents that must evolve as your SAP system, your business processes, and your organizational structure evolve. Outdated documentation is as risky, if not more risky, than no documentation at all, as it can lead to misinterpretations, incorrect actions, and system inconsistencies. Establish clear processes for review and update.

- Structured and navigable: It must be organized logically, with clear headings, subheadings, and an intuitive hierarchy that allows users to quickly navigate from high-level business process overviews down to granular configuration details or specific work instructions. Cross-referencing and hyperlinking are essential.

- Role-specific and audience-tailored: Different audiences have different needs.

- Executive leadership needs high-level strategic summaries and process impact overviews.

- Business process owners need detailed end-to-end process maps and decision rationale.

- End users need concise, task-oriented work instructions and troubleshooting guides.

- IT support teams need detailed configuration notes, technical specifications for integrations and customizations, and system architecture diagrams.

KEY TYPES OF DOCUMENTATION EVERY SAP CLIENT NEEDS: A COMPREHENSIVE VIEW

To ensure a robust and well-understood SAP environment, organizations need to prioritize the creation and maintenance of several critical types of documentation.

Business process documentation is the cornerstone. They map out the end-to-end "to be" business workflows (e.g., Order-to-Cash, Procure-to-Pay, Hire-to-Retire, Plan-to-Produce). They clarify how the business is meant to operate in the new SAP system, detailing steps, roles, responsibilities, decision points, and system interactions. This documentation ensures clarity and consistency across departments.

Configuration documentation is equally important, as it captures the intricate details of how SAP modules and functionalities are set up. Crucially, it documents not just what was configured (e.g., transaction codes, table entries) but the "why" behind each setting—the business rationale, the strategic decision it supports, and the implications of alternative choices. Without this context, future changes become a dangerous guessing game.

Master data documentation is critical to data integrity. This documentation defines how core data (customers, vendors, materials, chart of accounts) is structured, classified, governed, and maintained within SAP. It includes data definitions, data quality rules, data ownership, and data migration strategies. This prevents "garbage in, garbage out" scenarios and ensures data integrity.

Equally vital is security and roles documentation. This is vital for control and compliance. It details how user roles and authorizations are designed and implemented within SAP. It outlines the segregation of duties framework, preventing fraud and ensuring compliance with internal controls and external regulations. It specifies who has access to what, and why.

User guides and training materials serve a different purpose: They are written for the end users. They provide clear, step-by-step instructions on how to perform daily tasks within SAP. They can be role-specific (e.g., "Order Entry Clerk Guide") and often include screenshots, flowcharts, and FAQs. High-quality user guides ensure effective adoption and reduce reliance on constant support.

For every custom object developed—whether a report, interface, conversion, enhancement, form, or workflow (RICEFW)—customization documentation is essential. This includes technical specifications, functional design, testing procedures, and the specific business justification for the customization. This is particularly critical given the costs and risks of customizations.

Integration documentation ensures a clear understanding of how SAP interfaces with other internal or external systems (e.g., CRM, e-commerce platforms, banking systems, legacy applications). It details data flows, integration technologies, error handling, and reconciliation procedures. This is crucial for troubleshooting and evolving your integrated landscape.

Finally, test-case documentation provides long-term value well beyond the initial testing phase. Documented test cases (both for functional testing and user acceptance testing) serve as valuable assets for regression testing during future upgrades or enhancements. They ensure that new changes don't inadvertently break existing functionalities.

THE BUSINESS COST OF POOR DOCUMENTATION: THE UNSEEN DRAIN

Imagine this all-too-common, yet devastating, scenario: Your SAP project successfully goes live, perhaps with much fanfare. Three years later, your organization faces a critical need to make significant changes to the system—perhaps to accommodate new regulatory requirements, integrate an acquisition, or launch a new product line. However, none of the key individuals from the original project team or the external consultants are still at your company. The new internal team or a new consulting partner opens the SAP system, and what they find is a black box. There are no clear records of why certain configurations were made, how complex custom code interacts with standard processes, or what the original design rationale was.

This is not a hypothetical situation; it happens with alarming frequency. The result is a cascade of crippling consequences.

Massive delays and costly rework are often the first signs. Every necessary modification or enhancement becomes a laborious, expensive "guessing game" or a painful "rediscovery" mission. Hours, days, and weeks are wasted trying to decipher undocumented logic, leading to significant project delays and substantial additional consulting fees just to understand your own system.

Operational inefficiency and errors quickly follow. Without clear documentation, employees struggle to perform complex tasks, leading to increased errors, reliance on informal "tribal knowledge" (which is fragile), and the proliferation of inefficient manual work-arounds.

The risks extend to compliance as well. Audits become nightmares. If you cannot produce clear, documented evidence of your system's controls, processes, and data lineage, you risk regulatory noncompliance, substantial fines, and reputational damage.

Another consequence is consultant dependency and high support costs. Your organization remains perpetually dependent on highly expensive external consultants to "figure out" your undocumented system, draining your operational budget for basic support that should be handled internally.

Over time, poor documentation stifles innovation and agility. The inability to quickly understand and safely modify your SAP system means you cannot adapt rapidly to new market demands or competitive pressures. Your system becomes a static asset, hindering rather than enabling innovation.

In extreme cases, the sheer lack of documentation can lead to project paralysis. Major upgrades such as a move to S/4HANA or further transformations are almost impossible without the catastrophic decision of starting the entire implementation from scratch, effectively paying twice or thrice for your core ERP.

Poor documentation doesn't just create administrative inconvenience; it actively drains financial resources, undermines operational efficiency, elevates risk, and stifles strategic agility, making the long-term cost of initial shortcuts astronomically high.

MAKING DOCUMENTATION A SUPERPOWER IN YOUR PROJECT: EMBEDDING THE DISCIPLINE

To truly embed documentation as a strength and a superpower in your SAP implementation, it must be treated as a nonnegotiable strategic imperative, not an optional afterthought. This requires proactive planning, consistent discipline, and strong executive sponsorship.

Bake it into the plan from day one. Don't treat documentation as a separate, optional task. Explicitly assign dedicated time, sufficient resources (people and tools), and clear accountability for documentation creation and maintenance within every phase of your project plan, starting from the very first day. Integrate documentation tasks into daily work schedules, not just as a "big bang" at the end.

Ensure that consultants are accountable and contractually obligated. Make sure that your implementation partner's contract (statement of work) includes explicit, detailed requirements for documentation deliverables—their type, quality standards, format, and timing. Don't just assume they will provide it. Hold them accountable for delivering high-quality, comprehensive documentation as an integral part of their service, not an add-on. This includes formal knowledge-transfer sessions.

Additionally, create a robust governance process for documentation. Establish a clear governance framework for documentation that extends beyond go-live. This includes regular review cycles and updates for all critical documents (e.g., quarterly for processes, annually for configurations). This also includes clear ownership. Assign specific internal business or IT owners to each major documentation type or domain, making them accountable for its accuracy and currency. Implement strict version control to track changes, ensuring that you always have the latest, approved version to aid version control. Ensure any changes to the SAP system or business processes trigger a corresponding review and update of relevant documentation.

While consultants may generate the initial set of documents during the project, the long-term goal must be for documentation to shift from being consultant-created to client-owned. Your internal teams (business process owners, IT support, key users) should know where the documentation is, how to use it, and how to keep it current. This requires dedicated knowledge-transfer sessions and a culture of internal continuous improvement for owning the documentation internally.

Next, leveraging modern documentation tools can empower teams. Move beyond static Word documents and scattered spreadsheets. Invest in and leverage modern collaborative platforms, enterprise wikis, knowledge bases, or specialized document management systems that offer the following characteristics:

- Centralized repository and easy search functionality

- Version control and audit trails

- Role-based access control

- Collaboration features (e.g., commenting, coediting)

- Integration with other project management or IT service management tools

Consider tools that can automate some aspects of documentation (e.g., process mining tools for "as-is" process discovery, configuration documentation tools).

Lastly, integrating documentation with processes is not a separate activity. Embed documentation creation as a natural, seamless part of every project activity. When a design decision is made, document the rationale. When a configuration is completed, update the configuration document. When a process is tested, update the user guide with learnings. This makes documentation a continuous responsibility, not a separate, overwhelming task.

CASE STUDY: DOCUMENTATION AS A LIFELINE— THE AUDITABLE DIFFERENCE

Consider two chemical companies, both undergoing SAP implementations in similar regulatory environments but with vastly different approaches to documentation.

Company A: The Documentation Champion

This global chemical company invested heavily in meticulous documentation from day one. Their project plan included dedicated resources for documenting business processes, configuration decisions, master data definitions, and security roles. Consultants were contractually obliged to deliver this documentation to high standards, and internal business process owners actively participated in its creation and review. Three years after go-live, a stringent regulatory audit required immediate proof of how specific financial postings were configured and why, tracing them back to initial business decisions and compliance requirements. Because the company had treated documentation as a core strategic asset, they were able to produce clear, auditable records within days. This meticulousness saved them from significant fines, preserved their reputation, and allowed them to pass the audit with flying colors. The documentation served as an invaluable lifeline.

Company B: The Documentation Avoider

In stark contrast, Company B, in the same industry, viewed documentation as a burdensome overhead. Their project rushed to go-live, and documentation was largely an afterthought, consisting of scattered, incomplete files. When they faced a similar regulatory audit, they simply couldn't produce the required proof of configuration rationale or process controls. Their lack of documentation led to months of intensive, costly investigation, forcing them to reengage expensive original consultants just to "rediscover" their own system's setup. This delay and lack of transparency resulted in millions of dollars in penalties, a tarnished reputation, and significant operational disruption. In an even more severe consequence, their system became almost impossible to upgrade to the latest SAP version years later, due to the sheer unknown of what customizations and configurations existed without proper documentation.

The difference in outcomes was profound and directly attributable to one factor: One company treated documentation as an administrative chore to be avoided. The other recognized it as a strategic superpower, an essential investment in their long-term health, compliance, and agility.

In the complex narrative of every successful SAP project, documentation often plays the role of the unsung hero. It rarely commands headlines, receives public accolades, or sparks immediate excitement. Yet, quietly and persistently, it ensures organizational continuity, provides invaluable clarity, and builds unwavering confidence long after the grand go-live celebration has faded into memory. Leaders who truly grasp this profound truth transform documentation from a tedious chore into a powerful competitive advantage.

As you navigate your own SAP journey, shift your perspective. Don't passively ask, "Do we really need all this documentation?" Instead, proactively ask the empowering question: "How can we systematically make documentation our strategic superpower, embedding it so deeply that it fuels our long-term success?" The answer to that question could very well be the fundamental difference between an SAP implementation that slowly fades into confusion and technical debt and one that vigorously sustains, enables, and propels your business forward for decades to come.

You Can't Deploy SAP Without Changing People

Every SAP implementation arrives with a compelling suite of promises: the allure of streamlined processes, the power of integrated, real-time data, and the tantalizing prospect of unprecedented levels of efficiency and insight. Yet, despite these grand visions and massive investments, time and again, projects falter, stumble, or outright fail. The most common culprit isn't a technical glitch, a flawed configuration, or even a budget overrun; instead, it's a far more fundamental and often overlooked reality: People weren't ready, willing, or able to change the way they work.

The truth is stark and simple: You literally cannot deploy SAP without fundamentally changing people. SAP is not a plug-and-play tool that seamlessly slots into existing operations. Its very essence reshapes core business processes, alters decision-making hierarchies, redefines accountability, and even subtly, yet profoundly, shifts company culture. If you neglect or actively ignore this indispensable human side of the equation, no amount of technical brilliance, no matter how perfectly configured the system, will save the project from falling short of its transformative potential. The most sophisticated software in the world is useless if the people meant to use it either resist, misunderstand, or simply refuse to adapt.

Companies often make the critical mistake of underestimating just how deeply and pervasively an SAP transformation disrupts the daily routines and professional identities of their employees. It's not merely about learning a new software interface; it fundamentally alters not only what work gets done but how it gets done, who owns it, who is accountable, and what the new expectations are for performance and collaboration. This isn't just a technical shift; it's a profound emotional and psychological upheaval.

Consider the diverse impacts across different functional teams.

For finance teams, many have relied for decades on their meticulously crafted, often complex, personal spreadsheets for analysis, reconciliation, and reporting. Moving to automated postings, integrated general ledgers, and standardized reporting within SAP can feel like a significant loss of control over their data and their analytical process. They may fear losing their unique expertise or their perceived value to the organization. The shift demands a new level of trust in system-generated numbers and a different kind of analytical skill.

For sales professionals, the traditional emphasis might have been on speed and flexibility in closing deals, often relying on quick emails, informal agreements, or simple CRM systems. Suddenly, entering detailed, structured order data in SAP, adhering to standardized pricing conditions, and following rigid approval workflows can feel incredibly slower, cumbersome, and restrictive compared to their previous ad hoc methods. They may perceive the system as an impediment to their primary goal of selling rather than an enabler.

For operations (manufacturing/logistics) teams, employees on the factory floor or in the warehouse might be accustomed to highly localized, often intuitive, ad hoc problem-solving approaches to production scheduling, inventory management, or shipping. Following standardized, integrated workflows within SAP can feel restrictive, bureaucratic, and detached from the immediate realities of their physical environment. They may feel a loss of autonomy or that their practical expertise is being devalued by rigid system processes.

For customer service teams, moving from disparate systems where customer information is fragmented to a single, integrated SAP view can be powerful, but also daunting. Agents might struggle with new navigation, feel overwhelmed by the volume of integrated data, or resist standardized scripts if they feel it stifles their ability to provide personalized service.

The core of the challenge is that change is not just technical; it's deeply emotional. People resist SAP not primarily because they dislike technology but because it fundamentally disrupts familiar ways of working, challenges long-held habits, and often, without proactive management, triggers deep-seated fears about job security, competence, and a loss of professional identity. This cognitive load of learning new systems combined with the emotional toll of letting go of the familiar creates a powerful barrier to adoption.

THE DANGEROUS ILLUSION: "THE SYSTEM WILL FIX IT"

A common, yet incredibly dangerous, illusion held by some executives is the belief that SAP itself, merely by being implemented, will magically "fix" existing process problems or compel employees to adopt new, more efficient behaviors. This mindset views SAP as a self-implementing solution, a silver bullet that will automatically enforce discipline and drive transformation.

But the stark reality is that the system is only as effective as the people using it. If employees, either consciously or unconsciously, cling to old ways of working, resist new processes, or simply fail to understand the system's logic, SAP will rapidly devolve into a glorified reporting tool—a very expensive database—instead of becoming the integrated, transformative backbone of the business it was designed to be.

I've personally witnessed numerous companies invest millions, sometimes hundreds of millions, in meticulously configuring SAP, only to find that

- Employees consistently bypass the system for critical tasks, creating offline spreadsheets or unofficial databases.

- They resort to manual work-arounds that negate the automation and integration benefits of SAP.

- They input garbage data into the system because they don't understand its importance or the downstream impact of inaccuracies.

- They continue to operate in departmental silos, despite SAP's integrated design, because they resist cross-functional collaboration.

The result? The SAP system appears broken or ineffective, when in fact, the fundamental problem lies not with the technology but with the lack of widespread user adoption and behavioral change. This leads to wasted investment, continued inefficiency, and immense frustration for all stakeholders.

This is precisely why change management is not an optional add-on; it is the indispensable, hidden engine behind every truly successful SAP project. It's the proactive discipline that bridges the gap between technical implementation and human adoption, ensuring that the system's potential is actually realized through people.

LEADING CHANGE FROM THE TOP: THE UNWAVERING EXECUTIVE COMPASS

Change, particularly of the magnitude introduced by an SAP transformation, must unequivocally start at the leadership level. If executives treat SAP as "just an IT project"—delegating it entirely to the technology department, showing minimal personal engagement, or failing to articulate its strategic business value—then employees throughout the organization will inevitably adopt the same dismissive attitude. The project will be seen as a technical mandate, not a business imperative.

Leaders must actively and visibly lead the change by setting the tone and framing the narrative: From the very first announcement, leaders must consistently frame SAP as a fundamental business transformation, a strategic imperative for growth, efficiency, or competitive advantage, not merely an IT initiative. They must articulate the compelling "why" in a way that resonates with every employee.

Actions speak louder than words. Executives must visibly and consistently use the new SAP system themselves. This means abandoning old spreadsheets for SAP-generated reports, using the new dashboards, and adhering to the new processes. If leaders cling to old ways, frontline staff will perceive SAP as optional.

People need to hear not just what is happening (e.g., "go-live is next month") but consistently why it is happening (e.g., "SAP will give us real-time inventory data, so we can fulfill customer orders faster"). Communication should be frequent, multichannel, and tailored to different audiences, addressing their concerns and highlighting benefits relevant to their roles. Transparency about challenges builds trust.

Identify influential "super users" or "change champions" within each department. Empower them, train them deeply, and involve them in the design and testing phases. These champions become invaluable peer leaders who can support, mentor, and motivate their colleagues on the ground.

Publicly recognize and celebrate employees and teams who actively embrace the new system, adapt to new processes, and demonstrate its benefits in action. Showcase success stories to build momentum and reinforce positive behaviors. This creates positive reinforcement and encourages others to follow suit.

Leaders should walk the floor, engage with employees, listen to their feedback, and demonstrate empathy for their struggles. Their visible presence and genuine interest can significantly boost morale and commitment.

Without visible, unwavering, and consistent leadership buy-in and active participation, SAP will always feel like an imposed, optional burden to frontline staff rather than a shared journey toward a better future.

MANAGING RESISTANCE BEFORE IT BECOMES SABOTAGE: FROM OBSTACLE TO OPPORTUNITY

Resistance to change is a natural, predictable human response to disruption. It's not a sign of malice or incompetence; it's often a signal of fear, uncertainty, or a lack of understanding. The critical mistake leaders make is ignoring this resistance until it hardens into quiet sabotage—passive noncompliance, active work-arounds, or widespread cynicism that can cripple adoption. To manage it effectively, you need a proactive, empathetic approach:

- Identify stakeholders and their impact early: Before the project gains full momentum, systematically identify all key stakeholder groups who will be impacted by SAP. Who will SAP disrupt the most? Finance? Sales? Logistics? HR? Understand their current pain points, their perceived losses, and their potential gains. Categorize them as potential champions, resistors, or fence-sitters.

- Listen actively and empathetically: Create safe, open channels for employees to voice their concerns, fears, and frustrations. This includes town halls, anonymous surveys, focus groups, and direct conversations. People want to feel heard and validated. Often, initial resistance softens significantly once concerns are genuinely acknowledged and understood, even if not all demands can be met.

- Address fears transparently and proactively: Many fears are rooted in misinformation or anxiety about job security. Be transparent about how roles might evolve. Emphasize that while SAP may automate some transactional tasks, it often creates new opportunities for higher-value, more strategic work (e.g., data analysis, customer engagement, process optimization). Focus on upskilling.

- Provide safe spaces for learning and experimentation: Don't expect perfection from day one. Create environments where employees can experiment with the new system, make mistakes, ask "dumb" questions, and learn without fear of judgment or negative consequences. This might include dedicated training environments, sandboxes, or pilot programs. Allow time for practice and muscle memory to develop.

- Engage in cocreation, not just imposition: Involve key business users and process owners in the design and testing phases of SAP. When people feel they have contributed to shaping the new system, they develop a sense of ownership and are far more likely to champion its adoption.

- Resistance as a signal, not a threat: Reframe resistance within the project team. View it as valuable feedback—a signal that highlights potential gaps in your change strategy, areas of misunderstanding, or unforeseen operational challenges. Handled correctly, resistance can become a catalyst for refining your approach and improving the solution.

TRAINING AS TRANSFORMATION: BEYOND BUTTON CLICKS

One of the biggest, most common, and most costly mistakes I consistently observe in SAP projects is treating training as a mere afterthought—a checkbox activity to be completed in the final few weeks before go-live. This "just in time" approach to training is a recipe for disaster, leading to overwhelmed users, low proficiency, and widespread frustration.

Instead, to truly drive adoption and future-proof your system, generic, one-size-fits-all training is ineffective. Every employee needs training specifically tailored to the precise tasks, processes, and functionalities relevant to their individual role within the new SAP system. A finance clerk needs different training from a warehouse manager.

This means moving beyond simply showing "button clicks" or screen navigation. Employees need to practice realistic, end-to-end business scenarios—from creating a sales order to processing a customer return or managing a production run. This hands-on and scenario-based training builds confidence and muscle memory.

Learning a new ERP system is not a one-time event; training should be delivered in multiple, iterative phases, building familiarity and confidence over time. This includes initial foundational training, more advanced scenario-based training, and ongoing refreshers or training on new functionalities post-go-live.

Identify, train, and empower a network of highly knowledgeable and enthusiastic "super users" or "champions" within each department. These individuals become invaluable peer support, answering questions, providing informal coaching, and troubleshooting minor issues on the ground, significantly reducing the burden on central IT support.

Provide easily accessible, user-friendly documentation, quick reference guides, video tutorials, and a searchable knowledge base (as discussed in Chapter 15). This empowers employees to find answers independently.

Embed job aids, context-sensitive help, and guided tours directly within the SAP system where possible, providing immediate support at the point of need.

Good training doesn't just teach people how to use SAP; it helps them internalize why they should use SAP, how it benefits them personally, and how it contributes to the broader business goals. It builds competence, confidence, and ultimately, commitment.

CULTURE CHANGE: THE UNSEEN TRANSFORMATION

At its very core, deploying SAP is not merely a technological upgrade; it is a profound journey of culture change. The system, by its integrated nature, forces organizations to shift from old paradigms to new, more collaborative, and data-driven ways of working. This unseen transformation is often the most challenging, yet most impactful, aspect of an SAP project.

SAP breaks down traditional departmental silos. Data that was once "owned" by Finance or "managed" by Sales now becomes shared, integrated, and accessible across functions. This demands a shift from isolated departmental thinking to cross-functional collaboration and shared accountability for end-to-end processes.

Many organizations have historically relied on gut feelings, informal networks, or fragmented data from disparate systems for decision-making. SAP introduces a new level of data transparency and real-time insight. This requires a cultural shift toward data literacy, critical thinking about data, and a fundamental trust in system-generated information, moving away from subjective intuition.

In many legacy environments, "heroes" emerge who can fix broken processes with manual work-arounds or personal ingenuity. SAP, by contrast, thrives on standardization and discipline. The cultural shift is toward valuing adherence to standardized, efficient processes over individual "heroics" that create inconsistencies. Success is about following the system and leveraging its power, not working around it.

An integrated SAP system enables proactive management—forecasting demand, identifying supply chain bottlenecks, or anticipating financial trends. This requires a cultural shift from reactive firefighting to proactive planning and strategic foresight.

This cultural shift is profound and often deeply uncomfortable. It requires humility from leaders to challenge the status quo, openness from employees to embrace new ways of working, and unwavering persistence across the entire organization to embed these new behaviors and values. It's about rewiring the organizational DNA.

CASE STUDY: THE PROJECT THAT FAILED BECAUSE PEOPLE DIDN'T CHANGE

Consider the cautionary tale of a global retailer that invested heavily in an SAP S/4HANA rollout, aiming to centralize its operations and gain real-time visibility. From a technical perspective, the implementation was a triumph: The system went live on time, was technically flawless, and integrated seamlessly across all core functions—sales, procurement, finance, and supply chain.

However, a critical omission proved fatal: No one had invested adequately in strategic change management. There was minimal proactive communication about the "why" of the transformation beyond a generic "modernization" message. Training was rushed and generic, focusing on clicks rather than context. Key business users were not engaged in the design process.

Months after go-live, a disturbing pattern emerged: Employees in Sales and Procurement continued to track critical information (customer orders, inventory, supplier details) in their familiar Excel spreadsheets, completely bypassing the new SAP system. Why? They simply didn't trust SAP's numbers, didn't understand how their data entry impacted downstream processes, and didn't see the personal benefit of changing their habits. The system was technically perfect, but the people weren't using it.

Within a year, the company was paying consultants again, not to "fix" a broken system (because it wasn't broken technically) but to launch a massive, expensive "adoption program" to change the behavior of its people. The initial investment had largely gone to waste because the human side of transformation had been ignored.

PRACTICAL STEPS TO DRIVE HUMAN-CENTRIC CHANGE: A LEADER'S PLAYBOOK

To ensure your SAP project thrives through human adoption, leaders must proactively implement these practical steps:

- Frame SAP as a business transformation, not an IT Project: Consistently position the project as a strategic imperative for growth, efficiency, or competitive advantage. Articulate the "why" in compelling, business-focused terms that resonate with all employees.

- Engage leaders as visible role models: Executives and senior managers must actively use and endorse the new SAP system themselves. Their visible commitment and enthusiasm are contagious and signal to frontline staff that the change is nonnegotiable and valuable.

- Prioritize training and empower champions: Invest significantly in comprehensive, role-based, and scenario-driven training that occurs iteratively throughout the project. Identify, train, and empower internal "super users" or "change champions" to provide peer support and drive local adoption.

- Communicate relentlessly and empathetically: Develop a multi-channel communication plan that consistently repeats the "why," addresses concerns, celebrates milestones, and provides transparent updates. Tailor messages to different audiences, focusing on "what's in it for me."

- Manage resistance proactively and empathetically: Identify potential resistors early. Listen actively to their concerns, acknowledge their fears, and address them transparently. Create safe spaces for feedback and learning, transforming resistance into constructive dialogue.

- Celebrate wins and showcase success stories: Publicly recognize and celebrate individuals and teams who embrace the new system and demonstrate its benefits. Share success stories to build momentum, reinforce positive behaviors, and inspire others.

- Measure adoption, not just go-live: Beyond tracking system uptime, implement metrics to measure actual user adoption (e.g., number of active users, frequency of key transactions, reduction in manual work-arounds). This provides tangible evidence of behavioral change and value realization.

- Integrate change management into project governance: Ensure that change-management activities are not separate but are fully integrated into every phase of the project plan and are regularly reviewed by the steering committee. Make change management a core responsibility of project leadership.

Every SAP implementation, beneath its layers of code, configurations, and complex processes, is fundamentally a story about people. Technology may capture the headlines and consume the budget, but it is the willingness, ability, and enthusiasm of your people to adopt new ways of working that ultimately determines the outcome. If people don't change, the most technically perfect system will fail to deliver its promise. If people embrace new processes, new tools, and new ways of collaborating, SAP becomes far more than just software—it transforms into a powerful catalyst for profound business transformation.

That's why true future-proofing of your SAP investment is never just about code, configurations, or optimized processes; it is, first and foremost, about people. Because in the end, it's not SAP that transforms your business; it's your people, empowered, enabled, and inspired by SAP, who drive the real, lasting change.

Future-Proofing Your SAP Investment for Enduring Value

When companies embark on the monumental journey of an SAP implementation, an almost irresistible pressure often emerges: the intense desire to go live quickly. This urgency, driven by a confluence of factors, can overshadow virtually every other strategic consideration. Executives, eager to demonstrate progress and realize immediate returns on a massive investment, demand that the new system be in place with aggressive deadlines. Consulting partners, operating under contractual obligations and internal firm metrics, are equally driven to hit those go-live milestones, often incentivized by rapid completion. And employees, exhausted by the disruption of a major project, simply yearn to return to a state of "business as usual," hoping the new system will stabilize rapidly and allow them to resume their routines.

This collective rush, while understandable from a human and project management perspective, frequently leads to a series of shortsighted decisions that, in the long run, prove to be incredibly expensive and strategically detrimental. These compromises, often made in the name of speed, can have lasting consequences. One common shortcut is hardcoding processes just to meet immediate needs. Instead of designing flexible, adaptable workflows that leverage SAP's standard capabilities, teams might opt for quick, custom fixes that rigidly lock processes into their current-state molds. This avoids the immediate discomfort of process reengineering but creates future inflexibility.

Another compromise is ignoring scalability for future business models or growth. The focus becomes solely on making the system work for today's transaction volume, today's markets, and today's product lines, without adequately considering how it will handle exponential growth, entry into new geographies, or entirely new revenue streams (e.g., shifting from product sales to subscription services).

Teams also frequently cut corners on data governance and master data quality. The tedious, labor-intensive work of data cleansing, standardization, and establishing robust data governance is often minimized, deferred, or superficially addressed. This prioritizes a faster build over a clean, reliable data foundation, leading to "garbage in, garbage out" scenarios.

In many cases, the focus is placed on the speed of technical configuration rather than on strategic sustainability. Decisions are made to accelerate go-live, even if they introduce significant technical debt, limit future upgrades, compromise the system's long-term agility, or create undue reliance on specific individuals or external partners.

Finally, organizations often replicate legacy inefficiencies. Rather than seizing the golden opportunity to reengineer and optimize outdated, inefficient processes, teams might simply replicate existing workflows verbatim within the new SAP system to avoid internal friction and accelerate design. This digitizes existing chaos rather than transforming it.

The immediate result of this short-term thinking is often a system that appears functional and stable at go-live. However, beneath the surface, it's brittle and rigid. When the business inevitably evolves—as all businesses must to survive and thrive—those quick fixes and compromises begin to unravel. The system starts to crumble, resisting necessary change and forcing the organization into costly, painful rework. What initially seemed like a "faster path" or a "budget-saving shortcut" ends up being exponentially more expensive, more disruptive, and far more painful in the long run, eroding the very value the SAP investment was meant to create.

SAP IS A PLATFORM FOR THE FUTURE: AN EVOLVING ECOSYSTEM

It's crucial to fundamentally shift your perception of SAP. It isn't merely a static piece of software you install and then leave untouched; it's a sophisticated, dynamic platform designed to evolve continuously with your business. The truly critical question to ask during an SAP implementation isn't merely "Does this system work for us today?" but rather "Will it still effectively serve our strategic needs five, ten, or even fifteen years from now?"

The modern business landscape is in a constant state of flux, characterized by rapid technological advancements, shifting geopolitical dynamics, and evolving customer expectations. To remain competitive and relevant, your organization will inevitably need to adapt and grow in myriad ways. New markets will open up, requiring rapid expansion capabilities, localization of processes, and adherence to diverse regulatory frameworks. Your system must be able to scale globally with ease.

At the same time, global supply chains will continually reconfigure in response to geopolitical shifts, climate change, new sourcing strategies, or unforeseen disruptions. Your SAP must provide the flexibility to model and adapt these complex networks. Customers will increasingly expect seamless, personalized, and digital-first experiences across multiple channels, demanding new functionalities, real-time data access, and sophisticated integration points (e.g., e-commerce, mobile apps, IoT).

Regulations will tighten and proliferate across industries and geographies, requiring new compliance features, robust audit trails, and dynamic reporting capabilities. Meanwhile, technology itself will continue its relentless march forward, bringing transformative innovations like artificial intelligence (AI), machine learning (ML), robotic process automation, and the Internet of Things (IoT). These demand integrated data, flexible platforms, and the ability to consume new services.

In addition, business models will shift, moving from traditional product sales to subscription services, from one-time transactions to recurring revenue, or from traditional retail to complex omnichannel commerce. Your system must be able to support these new revenue streams and operational paradigms. Finally, mergers, acquisitions, and divestitures will occur, requiring rapid integration or separation of business units onto or from the core SAP platform.

If your SAP design is rigidly locked into today's processes, today's market conditions, and today's business model, you will inevitably be forced into costly, disruptive rework tomorrow. This means tearing out custom code, re-architecting data structures, or rebuilding entire process flows, effectively paying twice for functionality. If, however, your SAP system is built with foresight and a deliberate focus on future-proofing, it transforms into a resilient foundation that can gracefully absorb and adapt to change rather than resist it. It becomes an enabler of future growth, innovation, and strategic agility rather than a costly blocker.

DESIGNING FOR FLEXIBILITY: PRINCIPLES FOR A FUTURE-READY SYSTEM

How do you actually translate the concept of "building for tomorrow" into tangible design decisions within an SAP implementation? It begins with embedding core principles of flexibility, adaptability, and scalability into every layer of your system architecture and process design:

- Modular thinking and loosely coupled design: Configure processes and functionalities in a way that allows them to be adapted, enhanced, or even replaced without requiring a complete overhaul of the entire system. This means favoring standard SAP modules and functionalities and strategically using modern extension frameworks (like SAP Business Technology Platform—BTP) for side-by-side extensions rather than directly modifying core code. Changes in one area should have minimal ripple effects on others, promoting independent evolution.

- Data-centric approach and robust master data governance: Clean, standardized, and well-governed master data (customers, materials, vendors, financial accounts) is the fundamental fuel for future innovation. Design your data models with foresight, anticipating future analytical needs, AI/ML initiatives, and automation opportunities. Invest heavily in data quality and establish strong data governance processes that will sustain data integrity long term. Without clean, reliable data, future-proofing is impossible, as any advanced technology built on it will yield flawed results.

- Avoid over-customization (the roadblock of tomorrow): As discussed in Chapter 7, today's seemingly harmless customization is almost certainly tomorrow's expensive roadblock. Every piece of custom code, every bespoke report, and every deviation from standard SAP functionality creates technical debt that hinders future upgrades, limits innovation adoption, and increases maintenance costs. Stick to SAP's standard capabilities wherever possible, and rigorously justify any customization with a clear, long-term strategic value proposition. Prioritize process reengineering over technical modification.

- Global template mindset (scalability across geographies): If your organization operates in multiple countries or has ambitious plans for international expansion, design your SAP system with a global template mindset from the outset. This means defining core business processes, master data structures, and financial reporting standards that can be replicated and localized efficiently across different geographies. While some local flexibility may be necessary (e.g., for legal or tax requirements), the default should be global standardization to enable rapid, cost-effective rollouts and consistent operations worldwide.

- Cloud leverage and embracing continuous innovation: Embrace the power of SAP's cloud offerings (like SAP S/4HANA Public Cloud) or, for on-premise, plan for regular updates and upgrades. Cloud solutions, by their nature, often force standardization and provide continuous innovation through quarterly updates, new AI/ML features, and seamless integrations. This ensures your system remains current, secure, and future ready without the massive, disruptive, and costly upgrade projects that characterized older ERP versions.

- Design for scalability and performance: Build your SAP system with the explicit understanding that future growth will bring increased transaction volumes, a higher number of users, and potentially new types of data. Choose system structures (such as profit centers, plants, cost objects, organizational units, and technical infrastructure) that can easily accommodate significant future growth without requiring a complete redesign or performance bottlenecks. Consider future peak loads and data volumes.

- Focus on user experience (UX) and adaptability: A user-friendly and intuitive system is more likely to be adopted, adapted, and utilized effectively as business processes evolve. Leverage modern SAP Fiori interfaces and design principles that prioritize ease of use, reducing the learning curve for future employees and enabling quicker adoption of new functionalities or changes. A good UX ensures the system remains relevant and usable as your workforce changes.

- Build for advanced analytics and reporting flexibility: Design your data architecture, financial controlling elements (e.g., profit centers, cost centers), and other master data attributes with a clear understanding that future business intelligence and analytical needs will become increasingly sophisticated. This enables flexible reporting, predictive analytics, and advanced data-driven decision-making from day one, without requiring complex, separate data warehousing projects later.

CASE STUDY: THE COST OF BUILDING ONLY FOR TODAY—A LEGACY OF REWORK AND REGRET

Consider the experience of a large global consumer goods company that embarked on an SAP S/4HANA implementation with a heavy, almost exclusive, focus on replicating their current, highly specific workflows and organizational structures. Their primary goal was to get off their aging legacy system as quickly as possible, and they believed the fastest path was to customize SAP to match how they worked "yesterday." They invested heavily in bespoke ABAP code, custom reports, and unique process flows that mirrored their existing, often inefficient, operations across various regional silos.

At go-live, the system was technically stable and functional for their then current business model, primarily focused on traditional wholesale and brick-and-mortar retail. However, within just three years, their industry underwent a dramatic and irreversible shift toward direct-to-consumer e-commerce and complex omnichannel retailing. Their traditional business model was rapidly evolving, demanding real-time inventory visibility, personalized customer experiences, and seamless online-to-offline integration.

The customizations that once seemed harmless—designed only for their legacy wholesale processes—now became crippling blockers. Integrating with online marketplaces, supporting real-time inventory visibility for online orders, managing complex e-commerce fulfillment logic, and scaling customer service for direct consumer interactions required ripping out large, interconnected chunks of their existing SAP setup. They discovered that their custom code was not compatible with new standard SAP functionalities designed for e-commerce, and their rigid process flows could not support the agility required for online sales. They had to spend more money reworking and re-implementing their system post-go-live than they had spent on the entire original project. Their mistake was profound: They had built for yesterday, not tomorrow, creating a system that resisted the very evolution their business needed to survive and thrive in a new market paradigm.

THE INDISPENSABLE ROLE OF LEADERSHIP VISION: STEERING TOWARD THE FUTURE

Building for tomorrow requires more than just technical expertise; it demands profound leadership courage and unwavering strategic vision. Executives must actively resist the powerful, seductive temptations of short-term thinking and immediate gratification. One common pitfall is optimizing only for current-year ROI. A singular focus on immediate financial returns can lead to cutting corners that compromise long-term system health, agility, and strategic value. Leaders must balance short-term gains with long-term strategic investment.

Another challenge is the demand for customizations that feel familiar but add no future value. The comfort of replicating legacy processes often overrides the strategic imperative to standardize and simplify for future adaptability. Leaders must challenge these demands.

A third temptation is treating SAP as a one-off IT project instead of a living, evolving platform: This narrow mindset fails to allocate sufficient resources for ongoing maintenance, continuous improvement, and the adoption of new SAP innovations, leading to technological stagnation.

Finally, some leaders fail to articulate a clear, long-term business strategy: Without a compelling vision for where the company is heading, SAP becomes a tool without a purpose, unable to be designed for future needs.

Great leaders articulate a compelling vision of where the company is heading—its future markets, its evolving business models, its desired competitive advantages, and its long-term strategic goals. They then meticulously ensure that the SAP system is designed and implemented to robustly support that vision. The system should not only fit today's operational realities but also proactively anticipate and facilitate tomorrow's growth, innovation, and increasing complexity. This requires a strategic foresight that looks beyond the next quarter's earnings report and into the next decade of business evolution.

PRACTICAL GUIDELINES FOR BUILDING FUTURE-READY SAP: ACTIONABLE FORESIGHT

So, how can executive leaders and project teams translate the "build for tomorrow" philosophy into concrete, actionable steps during their SAP implementation?

- Anchor in long-term business strategy: Ensure that every major SAP design decision is explicitly aligned with your organization's long-term strategic goals, not just immediate process pain points or departmental demands. Use your strategic objectives as the ultimate tiebreaker in design debates and as the filter for all scope decisions.

- Invest heavily in master data quality and governance: Recognize that clean, standardized, and well-governed master data is the indispensable fuel for future innovations like AI, machine learning, predictive analytics, and advanced automation. Design your data models with future analytical and integration needs in mind. This is a nonnegotiable investment in your future capabilities.

- Favor standards over customization (relentlessly): SAP's standard functionality and industry solutions are continuously evolving. Staying as close as possible to standard functionality makes future upgrades seamless, reduces technical debt, lowers maintenance costs, and allows you to easily adopt new SAP innovations. Any customization must be rigorously justified by a clear, long-term strategic imperative and built using modern, upgradable extension frameworks (e.g., SAP BTP).

- Think global, act local (strategically): If your business has global aspirations or operations, build a core global template that standardizes common processes and data structures across entities. Allow controlled local flexibility only where absolutely necessary for legal compliance or critical competitive differentiation. This enables efficient global rollouts and consistent operations.

- Design for scalability and performance: Build your SAP system with the explicit understanding that future growth will bring increased transaction volumes, a higher number of users, and potentially new types of data. Choose system structures (such as profit centers, plants, cost objects, organizational units, and technical infrastructure) that can easily accommodate significant future growth without requiring a complete redesign or performance bottlenecks.

- Adopt a culture of continuous improvement and innovation: Treat go-live not as the end of a project but as the beginning of an ongoing journey. Establish governance models for regular process reviews, encourage employees to suggest enhancements, and create a pipeline for adopting new SAP functionalities and technologies. Embrace a mindset of perpetual optimization.

- Build internal capability for evolution: Invest in your internal SAP Center of Excellence, cross-training, and comprehensive documentation to ensure your team has the knowledge and skills to manage, support, and evolve the system independently, reducing reliance on external consultants for routine changes and enabling proactive innovation.

BALANCING TODAY VS. TOMORROW: THE ART OF STRATEGIC COMPROMISE

There is an inherent, natural tension in any SAP implementation between solving today's urgent operational problems and meticulously preparing for tomorrow's strategic demands. Go too far toward an imagined, distant future, and you risk overengineering, building unnecessary complexity, and failing to address immediate pain points effectively, leading to project fatigue and missed short-term goals. Focus exclusively on today and you inevitably set yourself up for costly, disruptive rework down the line, hindering future growth.

The answer lies in achieving a delicate, strategic balance. A truly future-ready SAP design doesn't ignore immediate needs; it solves them in a way that leaves ample room to grow and adapt. It's about avoiding decisions that paint your organization into a corner. This means prioritizing core, foundational elements such as master data quality, standardized core processes, and robust architecture that serve both today's operational stability and tomorrow's strategic growth. It also means implementing minimum viable products (MVPs) that deliver immediate, tangible value to address current pain points while simultaneously aligning perfectly with the long-term vision.

At the same time, it requires making incremental investments in future capabilities that don't derail current operations but position the business for future competitive advantage. Perhaps most importantly, it calls for the courage to say no to short-term fixes that create long-term technical debt, even if they offer immediate comfort. It's an art of strategic compromise, guided by a clear, shared vision for the future.

SAP is far more than just "getting live." It's about ensuring your business remains competitive, agile, and relevant in an ever-changing world. If you design your SAP system solely to solve today's problems, tomorrow's inevitable shifts in market, technology, or business model will quickly expose the cracks, turning your massive investment into a liability rather than an asset.

Remember this fundamental truth: Every shortcut you take today in an SAP implementation becomes tomorrow's expensive roadblock. Conversely, every thoughtful, disciplined decision you make today, every investment in foundational clarity and flexibility, becomes tomorrow's powerful competitive advantage. This foresight is what separates merely implementing software from truly transforming a business.

The companies that truly thrive with SAP are not the ones who boasted about "implementing fastest" or "going live cheapest." They are the visionary leaders who meticulously built robust, adaptable systems designed to last, to evolve, and to continuously support their business's strategic journey for years, even decades, into the future.

Building for tomorrow is not optional in an SAP transformation; it's the only way to ensure your investment is truly transformational and delivers enduring value.

16

Transformation Is Not for Sale

When companies make the momentous decision to implement SAP, there's a pervasive, almost subconscious, belief that they are, in essence, "buying transformation." The narrative often goes something like this: "We've identified our operational inefficiencies, our outdated systems, and our strategic gaps. Now, we will approve the substantial budget and sign the multimillion-dollar contracts with a leading software vendor and a top-tier consulting firm, and once the powerful new SAP software is live, our company will miraculously be transformed into a modern, agile, and highly efficient enterprise." This seemingly logical, yet profoundly flawed, mindset represents one of the most dangerous and costly myths in the entire landscape of enterprise technology. It's a comforting illusion, promising profound change without the inherent discomfort that true transformation demands.

The truth is stark and uncompromising: You can indeed buy SAP licenses. You can procure the expertise of highly skilled consultants. You can acquire cutting-edge methodologies, accelerators, industry templates, and even preconfigured industry solutions designed to streamline implementation. But what you absolutely, unequivocally cannot buy is transformation itself. Transformation is not a tangible product that can be purchased off a shelf, nor is it a simple line item in your budget that magically delivers results once paid for. Transformation is a dynamic, complex, and deeply personal lived process—one that demands unwavering leadership, profound personal ownership, and immense courage, not just from your executive team but from every single corner of your organization. It is an internal journey of adaptation and evolution, not an external acquisition.

THE MIRAGE OF "BUYING" CHANGE: UNPACKING VENDOR PROMISES

Vendors and consulting firms, in their understandable pursuit of securing lucrative contracts, often paint a compelling picture of transformation in their sales decks and executive presentations. They showcase impressive case studies, highlight revolutionary features, and articulate the potential for unprecedented efficiency. The allure is undeniable: Simply write a substantial check, flip a metaphorical switch (the go-live event), and then effortlessly reap the rewards of a modern, digitally enabled, highly efficient enterprise. This vision is seductive because it promises a relatively passive path to profound change, implying that the software itself is the primary, self-activating catalyst.

However, this is a dangerous mirage. Transformation doesn't happen merely by purchasing software; it happens when your leaders and your employees fundamentally change the way they think, the way they work, and the way they act. SAP is, without question, an incredibly powerful and sophisticated tool—a world-class platform capable of enabling unprecedented levels of integration, automation, and insight. It provides the robust infrastructure, the standardized processes, and the data backbone necessary for a modern enterprise. But it does not, and cannot, transform your business by itself. At best, it creates the essential conditions for transformation. It provides the framework, the data structure, and the process discipline that allow change to occur. Whether those conditions translate into a tangible, lived reality depends entirely on how effectively you, as a leader, guide your organization through the profound human, cultural, and operational shifts required.

If your people continue to cling to old habits, if your culture resists transparency and collaboration, or if your leaders fail to consistently enforce new ways of working, then the most advanced SAP system will remain an underutilized, expensive piece of shelfware. The software is the powerful engine; your leadership is the indispensable driver who charts the course, fuels the journey, and navigates the terrain. Without that active leadership, the engine remains idle or, worse, drives aimlessly.

TECHNOLOGY ENABLES, LEADERSHIP TRANSFORMS: THE ARCHITECT'S ROLE

To truly grasp this distinction, consider the analogy of a complex construction site. You can bring in the most brilliant architects to design the blueprint, the most skilled engineers to ensure structural integrity, and the most efficient construction crews equipped with the finest tools and machinery. You can purchase the highest quality steel, glass, and concrete. You can even hire a top-tier project manager to oversee the daily operations. But unless the owner of that project—the one with the ultimate vision for the building, its purpose, its aesthetic, and its future inhabitants—actively provides clear direction, makes decisive trade-offs, resolves conflicts, and commits to seeing the project through every challenge, the final result will fall far short of what was promised. The building might stand, but it won't be the functional, inspiring space envisioned; it might be a chaotic, compromised structure.

In the context of SAP projects, technology (the SAP software itself) is the raw material. Consultants are the expert builders, bringing the methodologies, technical skills, and man power to construct the system. But you, the client leader, are the indispensable architect of transformation. The strategic blueprint for your transformed business doesn't reside in a consultant's laptop or a vendor's code; it lives with you, in your vision for the company's future. You define the purpose, the desired functionality, and the aesthetic of the new operating model. Without your unwavering leadership—your strategic guidance, your commitment to process reengineering, your courage to challenge the status quo, and your relentless drive for adoption—even the most technically perfect SAP system will inevitably crumble under the weight of organizational inertia, internal resistance, or strategic misalignment. The system is the instrument; your leadership conducts the symphony, ensuring every section plays in harmony.

TRANSFORMATION REQUIRES PERSONAL OWNERSHIP: THE LEADER'S NONNEGOTIABLE ROLE

As we thoroughly discussed in Chapter 2, you simply cannot outsource your vision for an SAP implementation. The same fundamental truth applies, with even greater force, to the act of transformation itself. Transformation is not something you can delegate away; it demands profound, personal ownership from the top down. It requires leaders to be active participants, not just passive approvers.

As a leader, your direct, visible, and sustained ownership is the most powerful catalyst for change within your organization. This means you must

- Define and relentlessly communicate the "why" (the compelling purpose): Your employees need to see profound meaning in the disruption and effort of an SAP project, not just feel the burden of a mandate. You must articulate, with unwavering clarity and conviction, why this project is essential for the company's survival, growth, or competitive edge. Why now? Why this specific system? Why is it worth the personal and organizational effort? When your people understand the compelling purpose behind the change, they move from grudging compliance to genuine commitment. This isn't a one-time announcement; it's a continuous narrative that you must champion, adapting the message to resonate with different audiences across the organization.

- Model the desired change and behavior (lead by example): Actions speak infinitely louder than words. If you expect new behaviors, new processes, and a new mindset from your teams, you must embody them first. Leaders who continue to rely on old spreadsheets for critical data, bypass new SAP processes for convenience, or delegate all system interactions to their assistants send a silent, yet deafening, message: "This transformation is optional for me, so it's optional for you." Conversely, when employees see their executives actively using the new system, embracing new processes, and demonstrating a personal commitment to the new way of working, it signals that the change is real, important, and nonnegotiable. Your consistent actions are the most powerful form of communication and inspiration.

- Hold people accountable (with empathy and firmness): Transformation often dies in the "gray area" where new rules, new processes, or new expectations are not consistently enforced. You must hold individuals and teams accountable for adopting new behaviors, adhering to new processes, and leveraging the new system. This isn't about punitive measures; it's about clear expectations, consistent follow-through, and providing the necessary support. It requires courage to address resistance directly, but also empathy to understand the root causes of that resistance and provide necessary support (as explored in Chapter 13). When your people see your direct, unwavering ownership of the transformation—not just the purchase of a system—they are far more likely to follow, even when the path is challenging.

- Make the hard decisions (the power of "no"): As discussed in Chapter 12, transformation demands difficult choices. It requires saying no to beloved legacy quirks, no to unnecessary customizations, and no to short-term compromises that undermine long-term strategic goals. Your personal ownership means you are prepared to make these tough calls, even when they are unpopular, because you understand their critical impact on the ultimate success of the transformation. You become the ultimate arbiter, guided by the strategic North Star.

- Be present, engaged, and visible: Your ownership isn't just about making high-level declarations; it's about active, visible engagement throughout the project life cycle. This means regularly attending key steering committee meetings, participating in critical design reviews, engaging directly with project teams, and actively listening to feedback from frontline users. Your consistent presence signals commitment, empowers the project team, and builds trust across the organization. You become the face of the transformation.

CULTURE EATS SOFTWARE FOR BREAKFAST: THE UNSEEN FORCE

You may have heard the profound adage, often attributed to management guru Peter Drucker, that "culture eats strategy for breakfast." In the context of SAP projects, this truth is even more stark and immediately impactful: Culture eats software for breakfast.

You can invest hundreds of millions of dollars in the most advanced, perfectly configured SAP system, designed to integrate every function and streamline every process, but if your organizational culture is inherently resistant to change, if it fosters departmental silos and discourages cross-functional collaboration, if it hoards information rather than sharing it transparently, or if it simply treats SAP as "IT's project" rather than a shared business imperative, then no amount of money spent on technology will save you. The system will be undermined by the very human dynamics it was meant to improve, leading to low adoption, work-arounds, and a fundamental failure to realize value.

Conversely, if your organizational culture is characterized by openness to change, a collaborative spirit, a willingness to share information, a focus on continuous improvement, and a collective understanding that SAP is a tool for our business transformation, then even a challenging implementation can lead to extraordinary results. Such a culture can absorb disruptions, adapt to new ways of working, and leverage the system's power to its fullest potential, turning obstacles into opportunities.

Your organizational culture is not shaped by software; it is profoundly and continuously shaped by leadership. And that leadership, with its values, its behaviors, its priorities, and its communication, starts with you. Your unwavering commitment to fostering a culture that embraces change, values data-driven decision-making, and prioritizes enterprise-wide collaboration is far more impactful than any technical feature within SAP. It is the fertile ground upon which transformation can truly flourish.

THE TRUE INVESTMENT: BEYOND FINANCIAL OUTLAYS

When companies budget for an SAP implementation, the focus is almost exclusively on the financial outlay: software licenses, consulting fees, hardware, and internal resource allocation. They see the multimillion-dollar figure and assume that's the "hard part"—the primary hurdle to clear.

However, this perspective overlooks the true, often underestimated, investment required for transformation—an investment that cannot be quantified solely in monetary terms. Transformation doesn't just cost money; it costs your time, energy, courage, consistency, and patience.

Your personal, sustained involvement in steering the project, participating in key decisions, communicating the vision, and resolving conflicts is indispensable to its success. This is time diverted from other critical responsibilities, demanding a strategic re-prioritization of your schedule. It requires you to be present and engaged, not just an absentee landlord.

The immense emotional and intellectual energy it takes to keep people inspired when morale dips, to face resistance with patience, to navigate politics without losing sight of purpose, is something you only truly understand once you've lived it. Sustaining that energy means showing up—not once but every day—to listen, to encourage, to untangle problems no one else sees. It's the quiet, exhausting work of leadership that rarely gets recognized but defines whether a transformation endures. It's the energy of consistent championing, empathetic listening, and persistent problem-solving. This is the mental and emotional capital you expend daily.

The courage to reject convenient shortcuts, resist needless customizations, challenge outdated processes, and confront uncomfortable truths about readiness defines real leadership. It's the discipline to prioritize enduring strategic strength over temporary comfort or political ease. It's the courage to prioritize long-term strategic health over short-term comfort or political expediency.

The unwavering consistency required to reinforce the vision over and over again until it sticks, to model the desired behaviors day in and day out, and to ensure that every decision—big or small—aligns with the strategic direction is what ultimately determines whether transformation takes root or fades away. Inconsistency breeds confusion, erodes trust, and signals that the transformation is not truly serious.

It takes profound patience to recognize that deep, systemic change cannot be rushed—that setbacks will happen, adoption evolves over time, and people need space to adapt and grow. It's the patience to embrace progress in iterations while keeping a steady focus on the long-term vision. It's the patience to allow for iterative progress and learning, while maintaining a firm hand on the strategic rudder.

This is the real investment many leaders underestimate. They see the financial outlay and assume that's the hard part. The true, and often most challenging, investment is your sustained, personal leadership across the months and years of the transformation journey. This nonfinancial investment is what ultimately unlocks the financial ROI.

THE ILLUSION OF "GO-LIVE = TRANSFORMATION": THE BEGINNING, NOT THE END

One of the most pervasive and harmful myths in the SAP world is the belief that transformation is achieved at go-live. This misconception treats go-live as the ultimate finish line, the moment where the project concludes, the budget is closed, and all the promised benefits magically materialize. This leads to a premature sense of accomplishment and a dangerous relaxation of effort.

SAP go-live is not the end of the journey; it is merely the beginning. It's the moment your new operating model is born, and the real, continuous work of embedding the transformation, driving adoption, and realizing sustained value truly begins. The initial go-live is a technical milestone; transformation is a business outcome that unfolds over time.

True transformation is not a single event; it's a continuous process that shows up in tangible, sustained changes across your organization, long after the initial system is live. It's evident in the speed and quality of decision-making. True transformation is evident in how quickly, confidently, and accurately your leaders can make strategic decisions because your data is clean, integrated, and visible in real-time, replacing fragmented insights and manual reconciliation. It's also reflected in the improved experience for your customers. It's seen in faster order fulfillment, more accurate deliveries, personalized service, and seamless, consistent interactions that genuinely delight your customers and build loyalty.

For employees, transformation manifests in their ability to focus on higher-value activities, strategic analysis, and creative problem-solving rather than being bogged down in manual data entry, tedious reconciliation, or constant firefighting.

At the organizational level, it shows up in agility and adaptability—the enhanced ability to quickly adapt to new market conditions, launch new products, integrate acquisitions seamlessly, or pivot business models, leveraging SAP as a flexible, enabling platform.

Finally, true transformation is reflected in the culture of continuous improvement, where your organization actively seeks out and implements ongoing enhancements, optimizes processes, and leverages new SAP functionalities and technologies rather than treating the system as static.

If you mistakenly think of SAP as a product you simply bought, you'll measure success solely at go-live—a technical milestone. But if you truly understand SAP as a powerful platform for continuous business transformation, you'll measure success in how your organization fundamentally works differently, more effectively, and more strategically one, two, or five years down the line. The long-term value realization, not the initial deployment, is the ultimate metric of success.

LEADERSHIP AS THE DECIDING FACTOR: THE ULTIMATE DIFFERENTIATOR

In every single SAP project I've witnessed, without exception, the ultimate deciding factor for success or failure is not the sophistication of the software, not the reputation of the consulting partner, not even the size of the budget. It is, unequivocally, leadership.

Projects with committed, visionary, engaged, and courageous leadership consistently succeed—even when they encounter significant technical challenges, messy data, internal resistance, or unforeseen market shifts. The strength of leadership allows them to navigate obstacles, align stakeholders, make tough decisions, and maintain unwavering focus on the strategic outcomes.

Conversely, projects with absent, disengaged, inconsistent, or risk-averse leadership almost invariably fail to deliver their promise—even when the software is world class, the consultants are top tier, and the budget is ample. A vacuum of leadership creates a void that is quickly filled by inertia, internal conflict, short-term thinking, and ultimately, project paralysis.

The profound difference lies not in whether transformation was bought but whether transformation was genuinely led. Your leadership is the ultimate differentiator, the catalyst that transforms potential into reality.

TRANSFORMATION IS A VERB, NOT A NOUN: A CONTINUOUS JOURNEY

Transformation, in its truest sense, is not a static destination you arrive at and then check off a list; it is a continuous, dynamic verb. It is the ongoing act of leading your people through change, actively shaping your organizational culture to embrace new paradigms, and strategically leveraging your tools (like SAP) to continuously meet the evolving challenges and opportunities of tomorrow. It's a journey of perpetual adaptation and growth.

SAP gives you the powerful platform. Your consultants provide the specialized expertise and the building blocks. But only you, the client leader, can make transformation truly happen. It is an internal, human-driven process that you must champion, nurture, and sustain every single day. It's about building a resilient, adaptable enterprise that can thrive in an ever-changing world.

This book has walked you through the intricate phases and realities of SAP projects: from understanding the ideal process and avoiding early pitfalls to the critical importance of strategy over software, the dangers of customization, the power of data, the nuance of "best fit" practices, the necessity of empathy, the wisdom of direction over speed, and the superpower of documentation. It has laid bare the myths of and illuminated the paths to enduring success. I have lived these projects from the inside—as an SAP architect, a former SAP executive, and the SAP Gold Partner brought in when the stakes are high—and I wrote this book to give you the clarity and courage I wish every leadership team had in the room.

If you take away just one overarching message from this entire journey, let it be this: Transformation is not for sale. It starts, and ends, with you. Partners can amplify your effort, but they cannot lead in your place.

When you lead your SAP journey with unwavering vision, crystal clear clarity, and profound courage—when you personally own the "why" and model the "how"—SAP transcends its role as mere software. It becomes a robust, adaptable foundation for a truly transformed enterprise, capable of enduring and thriving in an ever-changing world. It becomes the digital backbone that enables your strategic ambitions. Lead that way and rooms calm, priorities sharpen, and execution accelerates without drama.

The consultants will eventually leave. The system will continuously evolve. The market will relentlessly shift. The only constant, the most powerful and enduring force in your organization's ability to adapt and succeed, is your leadership. If you want a steady hand beside you, I'm here. My team and I have helped organizations like yours separate signal from noise, align the room, and move fast without breaking what matters.

And in the end, that is precisely what makes transformation real, lasting, and profoundly impactful. It is your legacy. Own it with courage.

About the Author

SANJJEEV K. SINGH is a computer engineer, Harvard Business School alum, and member of the Forbes Technology Council. He is the founder and CEO of ASAR Digital, an SAP Gold Partner recognized for delivering some of the fastest and most cost-effective SAP Cloud ERP and SAP Customer Experience implementations in the industry.

With more than two decades of experience across the SAP ecosystem—as a consultant, as an SAP executive, and now as the leader of a high-growth implementation partner—Sanjjeev has helped organizations of all sizes unlock business value through digital transformation. His career has spanned hands-on solution architecture, global program leadership, and the creation of innovative, AI-driven tools that accelerate SAP adoption.

Sanjjeev is also an accomplished speaker and thought leader, frequently sharing his insights on enterprise transformation, ERP modernization, and the future of digital technology. He has presented to executives, industry groups, and technology leaders, earning a reputation for delivering clear, candid, and practical advice grounded in real-world experience. His contributions to the Forbes Technology Council and SAP community further reflect his commitment to shaping the conversation around business and technology transformation.

Through his writing, speaking, and leadership at ASAR Digital, Sanjjeev's mission is simple: to help companies move beyond seeing SAP as "just software" and instead harness it as a catalyst for lasting strategic transformation.